John Seiffertt and Donald C. Wunsch

Unified Computational Intelligence for Complex Systems

Adaptation, Learning, and Optimization, Volume 6

Series Editor-in-Chief

Meng-Hiot Lim
Nanyang Technological University, Singapore
E-mail: emhlim@ntu.edu.sg

Yew-Soon Ong
Nanyang Technological University, Singapore
E-mail: asysong@ntu.edu.sg

Further volumes of this series can be found on our homepage: springer.com

Vol. 1. Jingqiao Zhang and Arthur C. Sanderson
Adaptive Differential Evolution, 2009
ISBN 978-3-642-01526-7

Vol. 2. Yoel Tenne and Chi-Keong Goh (Eds.)
Computational Intelligence in
Expensive Optimization Problems, 2010
ISBN 978-3-642-10700-9

Vol. 3. Ying-ping Chen (Ed.)
Exploitation of Linkage Learning in Evolutionary Algorithms, 2010
ISBN 978-3-642-12833-2

Vol. 4. Anyong Qing and Ching Kwang Lee
Differential Evolution in Electromagnetics, 2010
ISBN 978-3-642-12868-4

Vol. 5. Ruhul A. Sarker and Tapabrata Ray (Eds.)
Agent-Based Evolutionary Search, 2010
ISBN 978-3-642-13424-1

Vol. 6. John Seiffertt and Donald C. Wunsch
Unified Computational Intelligence for Complex Systems, 2010
ISBN 978-3-642-03179-3

John Seiffertt and Donald C. Wunsch

Unified Computational Intelligence for Complex Systems

Springer

John Seiffertt
PhD Candidate
Department of Electrical and Computer Engineering
Missouri University of Science and Technology
Rolla, MO 65402
USA
E-mail: jes0b4@mst.edu

Dr. Donald Wunsch
Mark K Finley Missouri Distinguished Professor
Department of Electrical and Computer Engineering
Missouri University of Science and Technology
Rolla, MO 65402
USA
E-mail: dwunsch@mst.edu

ISBN 978-3-642-03179-3 e-ISBN 978-3-642-03180-9

DOI 10.1007/978-3-642-03180-9

Adaptation, Learning, and Optimization ISSN 1867-4534

Library of Congress Control Number: 2010928723

Typeset & Cover Design: Scientific Publishing Services Pvt. Ltd., Chennai, India.

Printed on acid-free paper

9 8 7 6 5 4 3 2 1

springer.com

Contents

Chapter 1
Introduction

1.1 The Need for Unified Computational Intelligence

Humanity's newfound ability to compute has granted us unprecedented acceleration of the rate of technological innovation, altered dramatically the structure of social engagement, and presented us with a chance to create a reality to match our imagination. From optimal control to bioinformatics to economic systems, the field of computational intelligence has proven itself a leader in maximizing what society stands to gain from its increasing supply of computing resources. Riding the bleeding edge of what computers are able to do, computational intelligence researchers find ways to increase the energy output from renewable sources, secure large-scale networks, stabilize power grids, control aircraft, detect skin cancer, find land mines, and even teach humans something new about games they have played for millennia. This book introduces ways to unify some key elements of the computational intelligence field in order to guide humanity's quest for a more perfect alignment between her most ancient dreams and her everyday life.

Computational intelligence encompasses a wide variety of techniques that allow computation to *learn*, to *adapt*, and to *seek*. That is, they may be designed to *learn* information without explicit programming regarding the nature of the content to be retained, they may be imbued with the functionality to *adapt* to maintain their course within a complex and unpredictably changing environment, and they may help us *seek* out truths about our own dynamics and lives through their inclusion in complex system modeling. These capabilities place our ability to compute in a category apart from our ability to erect suspension bridges, although both are products of technological advancement and reflect an increased understanding of our world. In this book, we show how to unify aspects of *learning* and *adaptation* within the computational intelligence framework. While a number of algorithms exist that fall under the umbrella of computational intelligence, with new ones added every year, all of them focus on the capabilities of *learning*, *adapting*, and helping us *seek*. So, the term *unified computational intelligence* relates not to the individual algorithms but to the underlying goals driving them. This book focuses on the computational intelligence areas of neural networks and dynamic programming, showing how to unify aspects of these areas

J. Seiffertt & D.C. Wunsch: Unified Computational Intell. for Complex Sys., ALO 6, pp. 1–17.
springerlink.com

to create new, more powerful, computational intelligence architectures to apply to new problem domains.

The first part of this book, Chapters 1 through 3, introduces an approach to unifying the ability of computational intelligence methods to *learn*. The second part, Chapters 4 through 6, discuss unification aimed at increasing the capability to *adapt*. Finally, the third part, Chapter 7, speaks to the computational spark allowing us to *seek*.

Before discussing an application that requires the use of a unified computational intelligence approach, a clarification of the usage of the word *unified* in comparison to the oft-used term *hybrid* is in order. As used in the literature, a hybrid computational intelligence technique is one that combines multiple algorithms into a single implementation. For example, a neural network trained with a particle swarm optimizer or an evolutionary algorithm that incorporates fuzzy logic in its fitness function may be classified as hybrid. The term *unified* refers to combinations along a different axis. For unified learning, a single architecture incorporates all three canonical learning modes into one unit so that signals from each learning mode can update and access the same content. This paradigm says, "You know these learning methods that are out there, segregated, each with their own algorithm? Well, here's the thing—they are actually the same algorithm." This is very different from what is meant by a hybrid algorithm, as a hybrid algorithm does not require the sharing of memory space. Also, the highly mathematical section of this book discusses unification in terms of input domains. There, the appropriate words are, "You know how there is one set of equations for continuous domains and another set for discrete domains? Well, here's the thing—these equations are actually the same thing." In this book, a framework is provided to deal cogently with multiple learning methods as subsets of unified learning in much the same sense that differential and difference equations are treated as special cases of the unified dynamic equations.

Further, it must be noted that *unified computational intelligence* means that a given algorithm seamlessly incorporates multiple learning modes that share weights and that may influence the already learned associations of the other modes, or that the algorithm will work on both discrete and continuous input signals using a single set of equations. In this way, these algorithms unify within the *learn* and *adapt* characteristics of computational intelligence approaches. *Unified computational intelligence* is not a term used to discuss an algorithm that can act as a neural network, a swarm intelligence system, and a fuzzy logic system all at once. The *unification* occurs at the "what is the algorithm doing" level, not at the "how is the algorithm classified" level. While this latter form of unification may be discussed within the purview of mathematical logic where all algorithms may be related, it is not within the scope of the current book.

What follows now is an example to help motivate the desire to develop such unified architectures. The example described is only one of many areas in which combining multiple learning methods can be fruitful. For example, Mohagheghi, Venayagamoorthy, and Harley (2007b) report on using multiple learning modes for a wide area controller for power systems, although their design is not "unified" in the sense described in this book. What it does, however, is give a powerful incentive to further develop algorithms that incorporate multiple learning modes, as the control of the power grid is one of the key control applications currently under investigation. Additionally, applications from finance are well suited for this approach.

We cover one such application in this field in the Future Work section. This example is a smart sensor application involving developing situation awareness for a force protection scenario. We provide here an overview and discussion of why unified learning is advantageous; full details may be found in Chapter 3. Figure 1.1 presents a graphical depiction of the problem. The practical considerations of this work include the need to develop field-deployable hardware capable of performing intelligent sensor fusion quickly, efficiently, and with minimum overhead.

An intelligent sensor fusion algorithm, like an intelligent creature, can make informed use of all three types of learning in this environment on the data set given. Certain paths may be pretrained prior to deployment, thus granting the human operators license to verify that the most obvious sensor patterns will be classified successfully. During operation, a reinforcement signal provided either by the environment or by the human operator acting off of the fusion algorithm's recommendations can adjust the current adaptive weight profile to curtail faulty clustering. Finally, in the absence of any external signal, the algorithm will learn in an unsupervised manner, comparing current inputs to what it already knows. In this way, all three learning methods are incorporated into a single application, providing a need for unified learning rather than a conglomeration of techniques pieced together in a computational sprawl.

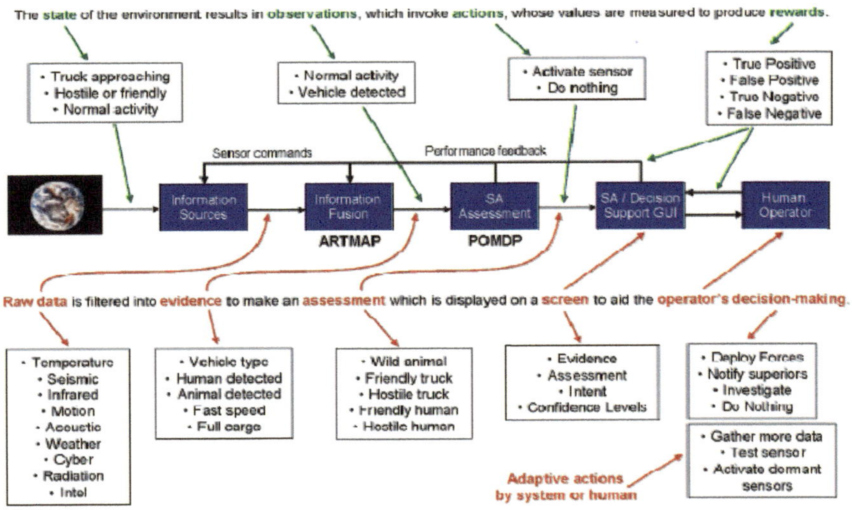

Fig. 1.1 An Application Using Unified Computational Intelligence

1.2 Contributions of This Work

The contributions of this book fall in the area of unified computational intelligence and encompass algorithm design, applications, theoretical developments, and the identification of new frontiers for multidisciplinary research.

Chapter 1 presents a new way to look at unified learning systems using a Markov Decision Processes framework. It also introduces new problem domains to which these unified approaches may be applied and defines the notion of unified computational intelligence.

Chapter 2 outlines an entirely novel Adaptive Resonance Theory-based unified learning architecture. From the motivation behind the algorithm to the design details to the extensions for future research, everything in this chapter is novel.

Chapter 3 contains an article that has appeared in the journal *Neural Networks* detailing an application of the architecture presented in Chapter 2.

Chapter 4 begins the theoretical component of the book. A new theorem in the time scales calculus is proven, and mathematics from scattered sources is brought together and organized for the first time.

Chapter 5 develops the theory of dynamic programming on time scales, one of the components of unified computational intelligence. This chapter contains new theorems regarding the nature of the dynamic programming algorithm and the Hamilton-Jacobi-Bellman equation in the time scales calculus. Also presented are new results from the area of quantum calculus. These results mark the first occurrence of the fields of time scales mathematics and dynamic programming being brought together.

Chapter 6 extends the time scales analysis to the domain of neural network learning, where the backpropagation algorithm is proven to hold in this new calculus as well as in its quantum calculus rendition. Additionally, the idea of an ordered derivative on time scales, a concept fundamental to the backpropagation algorithm, is introduced. The results in this chapter represent the first work to be published uniting times scales with neural network learning.

Chapter 7 discusses applications of computational intelligence in the emerging field of agent-based computational social science. It is increasingly important for researchers trying to make sense of complex economic, financial, and social systems to have as part of their technical vocabulary the language of computational intelligence. This chapter details how the approaches described in the book may be used in a setting outside the mainstream of engineering.

Altogether, this book introduces a new approach within the increasingly relevant field of computational intelligence and maps out new paths this research can take.

1.3 The Three Types of Machine Learning

1.3.1 Unsupervised Learning

Also called *clustering*, unsupervised learning refers to a situation in which an algorithm has no external guidance to focus its attention. When learning the mapping of inputs to clusters, it must rely solely on its own internal structure. For a full treatment of clustering, the reader is directed to Xu & Wunsch, 2008.

The type of unsupervised learning algorithm considered most thoroughly in this book is a neural network approach called Adaptive Resonance Theory (ART). Developed by Carpenter and Grossberg as a solution to the plasticity and stability dilemma, i.e., how adaptable (plastic) should a learning system be so that it does not suffer from catastrophic forgetting of previously-learned rules, ART can learn arbitrary input patterns in a stable, fast, and self-organizing way, thus overcoming the effect of learning instability that plagues many other competitive networks. ART is a learning theory hypothesizing that resonance in neural circuits can trigger fast learning.

ART exhibits theoretically rigorous properties desired by neuroscientists, which solved some of the major difficulties faced by modelers in the field. Chief among these properties is stability under incremental learning. In fact, it is this property that translates well to the computational domain and gives the ART1 clustering algorithm, the flavor of ART most faithful to the underlying differential equation model, its high status among unsupervised learning algorithm researchers. At its heart, the ART1 algorithm relies on calculating a fitness level between an input and available categories. In this way, it appears very much like the well-known *k-means algorithm*, although the number of categories is variable and grows dynamically as needed by the given data set.

What fundamentally differentiates ART1 from similar distance-based clustering algorithms is a second fitness calculation during which a given category can reject the inclusion of an input if the input does not meet the category's standards as governed by a single global parameter. Cognitively, this is modeling the brain's generation and storage of expectations in response to neuronal stimulation. The initial fitness, measuring the degree to which each input fits each of the established categories, is considered a short-term memory trace which excites a top-down expectation from long-term memory. Computationally, this second fitness calculation acts to tune the number of categories, and it may force the creation of new categories where a *k-means* styled algorithm would not, thus exhibiting stronger, more nuanced, classification potential. The ART1 algorithm has enjoyed great popularity in a number of practical application areas of engineering interest. Its chief drawback is the requirement that input vectors be binary. The ART2 algorithm was first proposed to get around this restriction, but the Fuzzy ART modification of ART1 now powers most of the new ART research and applications.

Fuzzy ART admits input vectors with elements in the range [0,1]. Typically, a sort of preprocessing called *complement coding* is applied to the input vectors, as well as any normalization required to map the data to the specified range. Fuzzy ART's core fitness equations take a different form than those of ART1, leveraging the mechanics of fuzzy logic to accommodate analogue data vectors. Fuzzy ART incorporates fuzzy set theory into ART and extends the ART family by being capable of learning stable recognition clusters in response to both binary and real-valued input patterns with either fast or slow learning.

The basic Fuzzy ART architecture consists of two-layer nodes or neurons, the feature representation field F_1, and the category representation field F_2, as shown in Figure 1. The neurons in layer F_1 are activated by the input pattern, while the prototypes of the formed clusters, represented by hyper-rectangles, are stored in layer F_2. The neurons in layer F_2 that are already being used as representations of input patterns are said to be committed. Correspondingly, the uncommitted neuron encodes no input patterns. The two layers are connected via adaptive weights, $\mathbf{W}_j$, emanating from node j in layer F_2. After layer F_2 is activated according to the winner-take-all competition between a certain number of committed neurons and one uncommitted neuron, an expectation is reflected in layer F_1 and compared with the input pattern. The orienting subsystem with the pre-specified vigilance parameter ρ ($0 \leq \rho \leq 1$) determines whether the expectation and the input pattern are closely matched. If the match meets the vigilance criterion, learning occurs and the weights are updated. This state is called resonance, which suggests the name of ART. On the other hand, if the vigilance criterion is not met, a reset signal is sent back to layer F_2 to shut off the current winning neuron for the entire duration of the presentation of this input pattern, and a new competition is performed among the remaining neurons. This new expectation is then projected into layer F_1, and this process repeats until the vigilance criterion is met. In the case in which an uncommitted neuron is selected for coding, a new uncommitted neuron is created to represent a potential new cluster. Researchers have concocted a wide variety of ART-based architectures by modifying the fitness equations to specialize them for a given problem domain.

For example, Gaussian ARTMAP uses the normal distribution to partition categories, with the relevant fitness equations incorporating the Gaussian kernel. This parametric statistical approach to ART was the first in what has become a rich field of study. Other parametric methods incorporate different probability distributions or allow for alternative preprocessing schemes based on statistics.

Other specializations of ART include ARTMAP-IC, which allows for input data to be inconsistently labeled and is shown to work well on medical databases; Ellipsoidal ARTMAP, which calculates elliptical category regions and produces superior results to methods based on hyper-rectangles in a number of problem domains; and a version of ART that uses category theory to better model the storage and organization of internal knowledge. Overall, Adaptive Resonance Theory enjoys much attention by those studying computational learning for both scientific and engineering purposes.

ART incorporates two steps: category choice and vigilance test. Let x be the input, w^j the weights associated with category j (this is really w_i^j where the weight is a vector of i elements, but this subscript is typically suppressed), and ρ be the vigilance.

In category choice, the degree of match is calculated:

$$\frac{|x \wedge w^j|}{|w^j|} \tag{1.1}$$

for each category j. In the ART calculations, $\wedge$ is the fuzzy AND operator, and $|x|$ represents the L_1-norm of x.

For the vigilance test, calculate

$$\frac{|x \wedge w^j|}{|x|} \tag{1.2}$$

and compare with ρ. The algorithm then cycles between category choice and the vigilance test until resonance occurs and weights update according to $w^j = w^j \wedge x$.

The relation given by Equation 2 calculates, in fuzzy logic terms, the percentage of w^j covered by x. See Figure 1.2 for a visual representation. The numerator tells us to what degree they overlap, and dividing then gives us a percentage. This is done so that the category choice reflects which category x is closest to. If one considers only the numerator, then the category choice value for the category that is identically equal to x (i.e., the perfect match) is the same as the category choice value for the uncommitted node of all **1**'s. The idea is to select the category to which the input fits just barely, and so the choice value is penalized for large w^j's. It makes no sense here to divide by $|x|$ because that quantity is the same for all categories; it would have no impact on which category is selected.

For the vigilance test, top-down expectations are checked. The category that won the competition in F2 is checked to see if it predicts that something like x ought to be the input pattern. In the fuzzy logic sense, this expectation checking is interpreted as follows:

$$\frac{|x \wedge w^j|}{|x|} \tag{1.3}$$

The algorithm now divides by $|x|$ to check what percentage of x is covered by w^j. Here, it makes no sense to divide by $|w^j|$ because it is not of concern how well w^j predicts itself. The check here, basically, is how many elements of the weights are less than elements of the inputs because w^j is everywhere larger than x this quantity is **1** and will not pass vigilance.

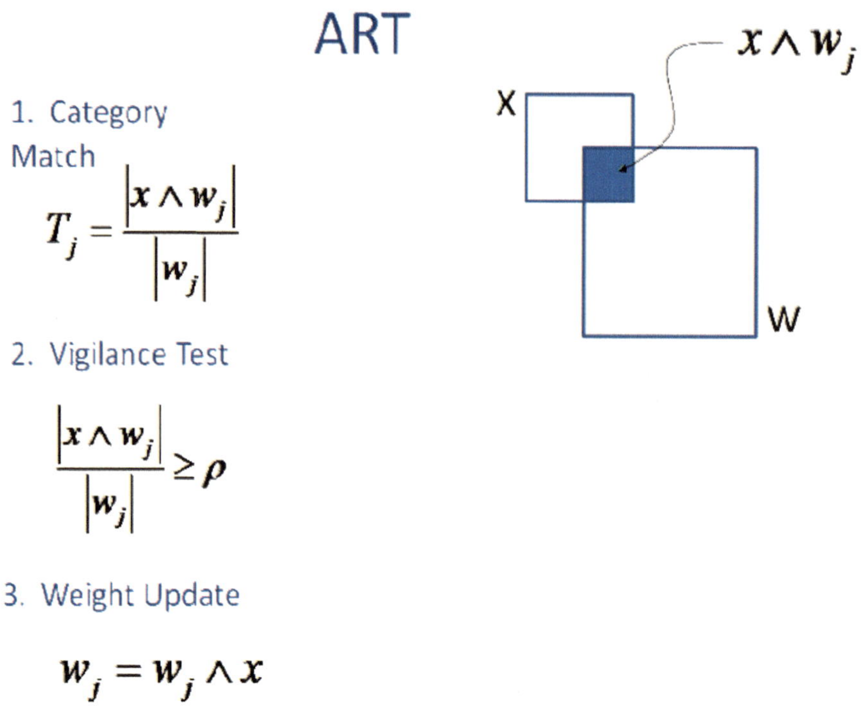

Fig. 1.2 An Adaptive Resonance Theory (ART) Classifier

This interplay between categories and inputs is the defining characteristic of ART systems.

1.3.2 Supervised Learning

Whereas in an unsupervised learning algorithm the system has no idea into what category an input *should* be placed, in a supervised learning problem this information is provided by the environment. It is the job of the system to match up the inputs with the given outputs, much as in a statistical regression problem. Typically, the supervised learning algorithm seeks to modify a list of adaptive weights in such a way as to minimize an error measure. In regression, these weights are the β_i coefficients. In a multi-layer perceptron neural network, these weights govern the movement of signals through the transfer functions at each layer. The neural network example is of principle importance and is studied in Chapter 6 in detail. In the development of the unified computational intelligence architecture of Chapters 2 and 3, however, the ARTMAP neural network is used as the supervised learning model.

1.3.3 Reinforcement Learning

Many fields, from animal learning theory to educational psychology, make use of the term *reinforcement learning* to mean a great variety of things. This book refers to a very specific mathematical definition of a problem type presented in Figure 1.3.

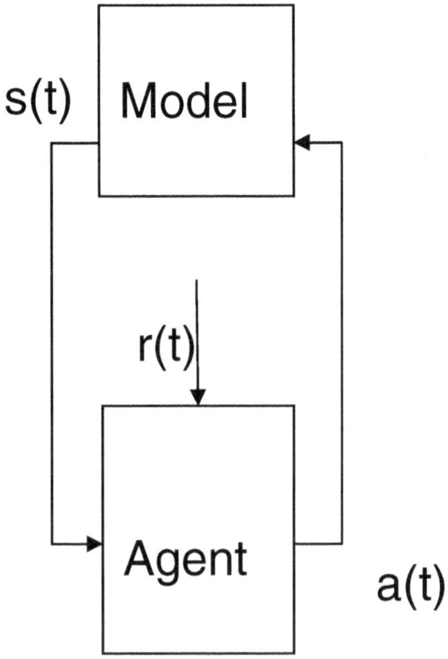

Fig. 1.3 Basic Reinforcement Learning Model Framework. Actions a(t), rewards r(t), and states s(t) are generated by the environment model and the agent controller

Some form of the Bellman equation is applied here to represent the agent's optimality criterion. It is important to understand that this literature hinges vitally on the notion of the agent as a maximizer of some utility function. In that way, there is much in the fields of economics and operations research that can usefully inform ADP theory (Werbos 2004.)

Barto and Sutton (1998) discuss a wide variety of solution methods for these problems. In particular, this chapter will focus on one solution method, a member of the TD-λ family of optimization algorithms (Sutton 1995), called Q-learning (Watkins 1989).

Note that the Q-learning algorithm, depicted in Figure 1.4, iteratively updates the value of each state-action pair. The appropriate modification is calculated based on the difference between the current and realized valuations, when maximized over all possible next actions. This is a key fact that sets up the more advanced techniques discussed in the next chapter.

Q-Learning Algorithm

1. Initialize Q(s,a)
2. Set t = 1
3. Initialize s
4. Set a = π(s), calculate s'
5. Update $Q(s,a) = Q(s,a) + \gamma[r(s') + \delta \max_{a'} Q(s',a') - Q(s,a)]$
6. Set s = s'
7. If s is not terminal, goto 4.
8. Increment t
9. If t is not equal to the maximum number of iterations, goto 3.

Fig. 1.4 The Q-Learning Algorithm

This algorithm utilizes a lookup table to store the Q-values for each state-action pair. As the scale of the simulation grows, the amount of memory required to catalogue these values can grow at a staggering rate.

Next, the generalization of the Q-learning algorithm to the artificial Higher Order Neural Network technique of Adaptive Critics is covered.

1.3.3.1 Approximate Dynamic Programming

A Widely-Used and Increasingly effective approach to solving problems of adaptation and learning in applied problems in engineering, science, and operations research is that of Approximate Dynamic Programming (ADP) (Si 2004, Bertsekas 1996). ADP techniques have been used successfully in applications ranging from helicopter flight control (Enns 2003), to automotive engine resource management (Javeherian 2004), to linear discrete-time game theory, a topic near and dear to the heart of many an economist (Al-Tamimi, 2007). As ADP techniques continue to enjoy favor as the approach of choice for large-scale, nonlinear, dynamic control problems under uncertainty, it becomes important for the computational economist to be aware of them. Approximate Dynamic Programming is a field grounded in mathematical rigor and full of social and biological inspiration that is being used as a unification tool among researchers in many fields.

This section overviews the structure of ADP. Markov Decision Processes are discussed first to introduce the core structural terminology of the field. Next, the Bellman Equation of Dynamic Programming, the true heart of ADP, is explained.

1.3.3.2 Markov Decision Processes

First, some terminology. The state of a system records all the salient details needed by the model. For an agent deciding how much of an asset to buy or sell, the modeler may set the state space to be a count of the current number of shares the agent is holding along with the current, stochastically generated dividend payment for the next time period. In the computational modeling of games such as Chess and Go, the relevant state would be the position of all the pieces currently on the board, and possibly the number of captured stones (for Go.) At each state, the agent has a choice of actions. (In a control application, where the agent is a power plant or some other complex operation to be optimally managed, the actions are called controls.) Our economic trading agent may buy or sell a certain number of shares, the totality of which entirely enumerates its possible actions. Returning to the game example, the entire range of legal moves constitute the action set for a given board configuration, or state. Each state nets the agent a level of reward. States that lead to desirable outcomes, as measured by some reasonable criteria, are assigned positive reward, while states that should be avoided are given negative reward. For example, the state arrived at after choosing the action that moves a Chess piece such that the opponent can place one's king in checkmate would generate a highly negative reward, while a winning Tic-tac-toe move would evolve the system to a state with high reward. The manner in which the agent proceeds from state to state through the choice of action is called the evolution of the system; it is governed stochastically through transition probabilities. The agent, upon buying a number of shares of a risky asset, finds itself in a new state. Part of the state's structure, the size of the agent's holdings, is under deterministic control. The stochastic dividend payment, however, evolves according to a statistical rule unknown to the agent. Therefore, the agent cannot know for certain to which state it will advance upon taking a certain action. Instead, the next states constitute a probability distribution described by the transition probability matrix. To contrast, the evolution is completely deterministic in Chess and Go, as no randomness is involved.

The way the state has been defined, as embodying all necessary information to calculate the future system evolution, allows the use of a mathematical Markov chain to model the system dynamics. Any such system, said to satisfy the Markov Property, can be analyzed with the following techniques. In practice, systems of interest often have a degree of error in the state representation, or some other influx of imperfect information, and therefore do not technically fulfill the Markov Property. However, approximation techniques for these situations abound, and the careful researcher still can make appropriate use of Markov chain modeling in many cases. For a more thorough analysis of such cases, see Sutton and Barto (1998).

A Markov Decision Process (MDP) model is one in which Markov chains are used to analyze an agent's sequential decision-making ability. In MDP terminology, the agent calculates a policy, an assignment of an action to every possible state. The goal is to find an optimal policy, given some reasonable criterion for optimality. An MDP consists of the components previously defined: states, actions, rewards, and transition probabilities. The time scale under consideration is also important. Discrete MDPs typically evolve along the positive integers, while continuous MDPs are defined on the non-negative real numbers. Other time scales are feasible for constructing MDPs. See the book by Bohner and Peterson (2001) for a more rigorous mathematical presentation of time scales.

MDPs have been studied and applied extensively in such areas as inventory management (Arrow 1958), behavioral biology (Kelly 1993), and medical diagnostic testing (Fakih 2006). Standard solution techniques are available and well understood (Puterman 1994). Solutions consist of an optimal policy for the agent to follow in order to maximize some measure of utility, typically infinite horizon expected reward.

It is not always the case that a system can be adequately expressed as a standard MDP. When the state information is not fully available to the agent, then the model must be supplemented with a probabilistic description of the current state, called a belief space. An MDP under this addition becomes a Partially Observable Markov Decision Process (POMDP). A classic POMDP example involves an agent deciding which of two doors to open. Behind one is a tiger, and behind the other is a lovely prince or princess ready to sweep the agent off its feet. In a straight MDP, the agent would have access to the transition probabilities for the two states and would be able to calculate which door is most likely to contain the desired result. In the POMDP formulation, however, the agent does not have access to such information. Instead, the agent receives observations, such as hearing the tiger growl, that combine to form a Bayesian approach to solving the optimal policy. POMDPs have demonstrated an ability to model a richer set of systems than the pure MDP formulation. For example, POMDPs have been used in dynamic price modeling when the exact demand faced by the vendor is unknown (Aviv 2005). When the demand at each period is known, an MDP can be used to calculate the best policy under expected reward criteria. But, when faced with an unknown state element, the agent must refer to observations such as historical marketing data to help make its decision.

Standard solution methods for POMDPs work only on specific frameworks and require significant computational capability to implement. To avoid these problems, it is common to use a technique such as a Bayesian Filter to transform a POMDP into an MDP once the observations key the agent's belief space to a sufficient degree. The solution techniques for MDPs then can be applied to the POMDP and the optimal policy calculated.

The next section provides the mathematical formulation of the task of the economic agent—the maximization of a particular optimality criterion.

1.3.3.3 The Bellman Equation

Consider an economic agent modeled with a finite set of states s, actions a, rewards $r(s)$, and transition probabilities $P(s, a)$ in a discrete time scale defined to be the positive integers. In order to calculate the agent's optimal policy, some utility function must be maximized. In the core Approximate Dynamic Programming paradigm, the function to be maximized is the Bellman equation:

$$J(s) = r(s) + \gamma \sum_{s'} P(s', a) J(s', a) \tag{1.4}$$

This is the discounted expected reward optimality criterion. In this equation, $J(s)$ represents the current value of a given state, s' signifies the next-states, and a discount factor $\gamma$is applied to the future rewards. This equation is to be maximized over all actions. Note that this is a special case of the full Hamilton-Jacobi-Bellman equation studied in Chapter 5.

The Bellman equation states that the current value of a state is equal to the immediate reward of taking an action plus the discounted future reward that accrues from that state. Other optimality criteria are possible to account for infinite horizon or nondiscounted models. The task of ADP is to solve this equation.

One standard solution algorithm is that of backwards induction. Other approaches include value and policy iteration. The interested reader is directed to Puterman (1994) and similar texts for further details on these and other optimization techniques. The solution method to be discussed in this chapter is found in the next section.

1.3.3.4 Heuristic Dynamic Programming

Q-learning is robust and has been shown to work quite well in a large number of problem domains, including being the base of the temporal difference approach at the center of a computational agent which, without any exogenously provided understanding of the rules of Backgammon, learned to perform at the master level and which was able to teach new strategies to arguably the world's oldest game to champion-level players (Tesauro 1994). However, its reliance on a lookup table to store values is a severe limitation. Generalizations of Q-learning, falling under the heading of Heuristic Dynamic Programming (HDP), replace the Q-table with a multi-layer neural network function approximator. Another generalization of Q-learning, dubbed Z-learning, involving a variable transformation to linearize the underlying MDP formulation, has been introduced (Todorov, 2007).

The diagram for HDP, the simplest of the class of architectures broadly known as Adaptive Critic Designs (Werbos, 1992, Prokhorov, 1997), is presented in Figure 1.5.

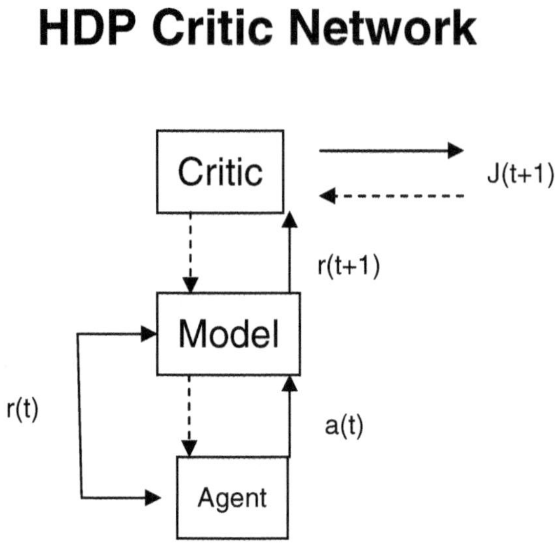

Fig. 1.5 Basic Adaptive Critic Design. J(t) is the value function being approximated, r(t) is the reward, and a(t) is the action control signal. The critic evaluates the agent's choice, modifying its adaptive weights in response to the chosen actions

The Adaptive Critic architecture, in essence, translates a reinforcement learning problem into a supervised learning problem. This is beneficial because much is known about solving supervised learning problems. The critic network learns the value function, and error between the current J-function value and the J-function value in the next time step is backpropagated through the network (Werbos, 1990).

Adaptive Critic architectures have found many application areas, including missile control (Han, 2002 and Chuan-Kai, 2005), fed-batch biochemical process optimization (Iyer, 2001), intelligent engine control (Kulkarni, 2003), multimachine power system neurocontrol (Mohagheghi, 2007), and even the management of a beaver population to prevent nuisance to humans (Padhi, 2006). The promise of finding rewarding application of these techniques in the fields of computational economics and finance is too alluring to ignore.

1.4 A Unified Approach

Based on the Markov Decision Process framework, a unified approach to the three learning modes can be established. In this unified view, unsupervised and supervised learning modes are seen as the extreme ends on a continuum of reinforcement learning. Unsupervised learning corresponds to the absence of any reinforcement, and supervised learning corresponds to the presence of perfect reinforcement.

First, consider the MDP model of a reinforcement learning agent. In this case, there are states representing salient features of the environment, and there is a mechanism to transition from one state to another in the presence of a control or action. The value of the reinforcement will help to determine if the chosen action evolved the state of the environment into a more or less valuable position. In the context of an autonomous robot navigating a maze or an intelligent controller minimizing a cost function, the structure of the reinforcement learning model is intuitive. Note that the *critic* is the unit that processes reinforcement, and the *actor* is the unit that calculates the action.

For a supervised learning problem, however, such intuition takes more effort to establish. In this case, the environment consists solely of the inputs to be learned. Regardless of the selected control, the environment will always transition to the next input on the list. However, the reinforcement signal is not to be interpreted as a measure of value; rather, it is communicating what the correct action ought to be, which requires the supervised learning critic to behave differently than the reinforcement learning critic. Where the RL critic may hold a value function that chronicles the appropriateness (as determined by the environment) of each state-action pair, the SL critic tracks no such information. Rather, it is a function that will update adaptive weights to better coordinate the input with the actual signal received. Many RL critics do this as well when they backpropagate an error through the neural-network based actor, but in the SL case this is more explicit—the error signal to be backpropagated is generated not from the critic's value table but from an error function, typically a least squared error measure between the input and the desired output (found in the signal from the environment).

In this way, the supervised learning mode emerges as a subset of reinforcement learning. The state is now simply an input vector, the transition probability matrix gives the next input vector with probability 1, the reinforcement signal from the environment contains the desired action value (target), and the action itself represents the target vector.

For the UL case, there is no reinforcement signal at all. The environment will generate a next state, which, as in SL, is interpreted to be the next input vector on the list, but the reinforcement signal is not present.

Notice that in the UL operation, the critic is dormant; without a signal from the environment for it to process, it may as well not exist. In order to see this as a subset of more general learning, however, it is useful to view the critic as existing but simply watching the action around it take place rather than taking an active role. In this case, all the internal machinery determining to which category an input is to be assigned is located within the actor. And, as in the SL case, the transition matrix for the input is degenerate, and the system always transitions to the next input vector as the "state."

These ideas are illustrated using the following equation model based on the Q-learning algorithm. Consider the following basic update rule:

$$Q(s_t, a_t) \leftarrow Q(s_t, a_t) + \alpha[r_{t+1} + V - Q(s_t, a_t)] \tag{1.5}$$

This is the basic Q-learning formula, with the V component equal to the internal operation of the critic. If the idea of the critic is expanded to refer to whatever internal calculation the algorithm performs, then we may use this equation to perform both supervised and unsupervised learning in addition to reinforcement learning. Of course, for both supervised and unsupervised learning methods, the reinforcement term r_{t+1} would be zero, but the value of V would depend on the dynamics of the particular approach being utilized. So, for a basic supervised algorithm, the V would represent least-squared error, and for basic unsupervised learning, the V would equal the fitness calculation between the input and any given category. More advanced algorithms could have V representing the ARTMAP or ART dynamics or a fuzzy c-means fitness, etc. A single update equation can be interpreted, through the activity of the critic, to perform any of the three learning modes. This is not to say that Q-learning leads to the most effective approach to supervised or unsupervised learning, only that this is the beginning of a single framework within which to discuss the three types. This framework will form the basis of the algorithms presented in Chapter 2, which show integrated learning in action.

1.5 Future Work

Extensions of the work presented in this book will cover new applications, more robust synthesis of the learning modes, and further convergence of discrete and continuous signals within a single theoretical framework.

In the application domain, there are many problems in social science in which an agent equipped with multiple learning modes may prosper. For example, a commonly studied computational economics situation is one in which a collection of heterogeneous agents must analyze an input signal that gives information about the future value of a risky asset. These agents must decide how to allocate resources between the risky asset and the risk-free asset so as to maximize payoff. This problem requires the agents to learn a match between the information signal and the future performance of the risky asset. Simple multi-layer perceptron neural network architectures have been employed successfully in approximating the functional relationship between these two signals of interest. However, in a more realistic application environment, the information signal would be expected to be intermittent, requiring the agent to process in an unsupervised mode while access to the information stream was limited. In this way, the use of a unified learning method, where the content learned in supervised and unsupervised modes would be mixed together within a single memory, may be the correct model of actual economic agents. It helps that the ART-based neural networks are themselves designed specifically to describe how biological neural and cognitive systems operate.

Furthermore, in a more robust application requiring portfolio balancing, an agent would be faced not with a single information source but with multiples. Advances in the theory of unified computational intelligence applied to the learning modes can generate algorithms that rate different supervisory signals via the reinforcement signals. When a supervisor indicates a match that subsequently generates poor reinforcement results from the environment, then the sagacity of that supervisor will be in doubt. In this way, an extension of the unified computational intelligence learning neural network presented in this book may be able to process multiple supervisory sources. This would be an entirely new problem domain for machine learning research.

In general, the development of new learning algorithms that are capable of synthesizing different sorts of feedback signals is a much needed extension of this book. The idea that multiple signals may impact the same stored memory is one not yet represented in the literature. Further theorems and applications regarding the interplay among competing learning signals are needed.

Theoretical extension of the simple MDP model presented in this book is needed. More theorems concerning the view of supervised and unsupervised modes as the extreme versions of the reinforcement mode would help guide the development of further algorithms for applications.

On the dynamic programming front, further analysis of the unification of discrete and continuous signals is needed. The time scales calculus is still an emerging area with relatively few active researchers worldwide. As more and more of the fundamental applied mathematics gets formulated in this calculus (multiple valued Taylor series, n-variable chain rules, variational calculus, functional analysis, and nonlinear systems theory, to name a few), new results on the control front will become available. Single algorithms capable of operating in discrete or continuous time will be available and may prove worthwhile in applications. In 20 years, aspiring control theorists may have to know a thing or two about the time scales calculus and dynamic equations.

Extensions of the quantum calculus and dynamic programming are of value as well. A quantum dynamic programming can be devised using the state space representations of quantum mechanics instead of the ones in classical mechanics. This approach may allow decision theory under uncertainty to be formulated in a new and useful way.

Finally, the area of agent-based modeling of social systems in general is still in its infancy. As more computational results contribute to an understanding of complex systems to the same degree as has traditional analytic mathematics, more scientists will understand the power behind these methods and begin using them with wider acceptance. Computational intelligence techniques, themselves inspired in a way by biological systems, extend the possibilities of computation and may fuse with these agent-based models in a vital way to help advance both basic and applied research into the governing dynamics of social and economic systems.

Chapter 2
The Unified Art Architecture

2.1 Introduction

The previous chapter introduced the idea of unified computational intelligence and discussed its implications for a learning machine. This chapter presents the design unified ART architecture, and Chapter 3 contains an application implementing this design.

Based on the Adaptive Resonance Theory neural network family, the Unified ART architecture is capable of seamlessly switching among unsupervised, supervised, and reinforcement learning without the aid of a governing control signal. Sections 2.2 through 2.4 motivate and outline the basic operation of the system as applied in Chapter 3 and in Section 2.4 presents an extension of the system.

2.2 Motivation

The distinguishing characteristic of ART systems is their use of what this book terms *permissive clustering*. That is, while most clustering algorithms such as K-means calculate a fitness value and assign each input to its fittest category, in ART systems the category itself is able to reject an input from being placed in it if that input is too dissimilar from what the category would "expect" one of its members to look like. Physiologically, this corresponds to the "top-down expectation" calculation performed by a biological computational unit such as a visual cortex when presented with stimuli from the sensory unit. Algorithmically, this adds to the computation a second step following the fitness calculation. During this step, a second test is run. If the input passes this test, then it is allowed to be placed in the category corresponding to the prevailing high fitness; otherwise, the input must proceed to the next fittest category and present itself for testing according to that category's criteria.

In the ART literature, this second step is called the *vigilance test* and is controlled by a single scalar called the *vigilance parameter*, typically denoted by ρ. It is crucial that ART researchers understand the importance and uniqueness of this second level of permissions. It is so easy to get carried away by attempts to

J. Seiffertt & D.C. Wunsch: Unified Computational Intell. for Complex Sys., ALO 6, pp. 19–32.
springerlink.com

streamline the computational algorithm or generate tractable mathematics that a researcher may end up working on an architecture that does not actually belong to the ART lineage. Many worthy algorithms purport to be versions of ART but do away with the vigilance test entirely. While these algorithms may be fine solutions to the engineering problems for which they were developed, it is important to keep in mind the core principles of ART when designing these architectures. In this way, a connection with the biological structures underlying the behavior of ART systems can be maintained, and the strong ties between computational intelligence and nature are kept alive. While the resulting system need not be a slave to what the natural world provides, much insight may be gained by using its revelations to help engineer solutions.

The three learning modes of supervised, unsupervised, and reinforcement learning receive the most attention in the literature. The unified architecture is a seamless integration of these three learning modes in that (1) it uses a single set of adaptive weights to process all three modes, (2) it determines when to use which learning mode without the intervention of an operator, and (3) the various signals driving the learning modes all interact with each other at the base learning level. Similar research on integrating reinforcement learning with the existing ARTMAP systems is not fully integrated in this sense. They instead use the ART classifier to cluster inputs and then process these clusters using a variety of reinforcement learning methods, most often temporal difference methods. While the usefulness of clustering inputs is undisputed in some problem domains, the resulting architecture cannot be said to be fully integrated in the sense of an ART system. This book presents the first architecture that recognizes the unique features of ART and uses them to create a fully integrated learning machine. In this way, this contribution is the first unified learning neural network architecture to appear in the literature.

To motivate the algorithm presented in the next subsection, the development of ART from an unsupervised algorithm to its incarnation as ARTMAP, a supervised learning tool, is outlined. It is important to see that a new learning mode added to an ART network must interact with the vigilance test. ARTMAP will override a successfully passed vigilance test if the supervisory signal so demands. It is this interaction between the new learning mode—the presence of the supervisor—and the vigilance test that places ARTMAP firmly in the ART camp. A system integrating reinforcement learning into ART must interact with the vigilance test. The RL signal is capable of causing a category reset in the ART module, thus influencing the nature of the stored prototypes and the structure of the intelligent network. This feature more faithfully emulates biological systems and also provides for a more robust interaction among the three learning types.

2.3 Block Diagram

Figure 2.1 shows the block diagram for the unified ART architecture.

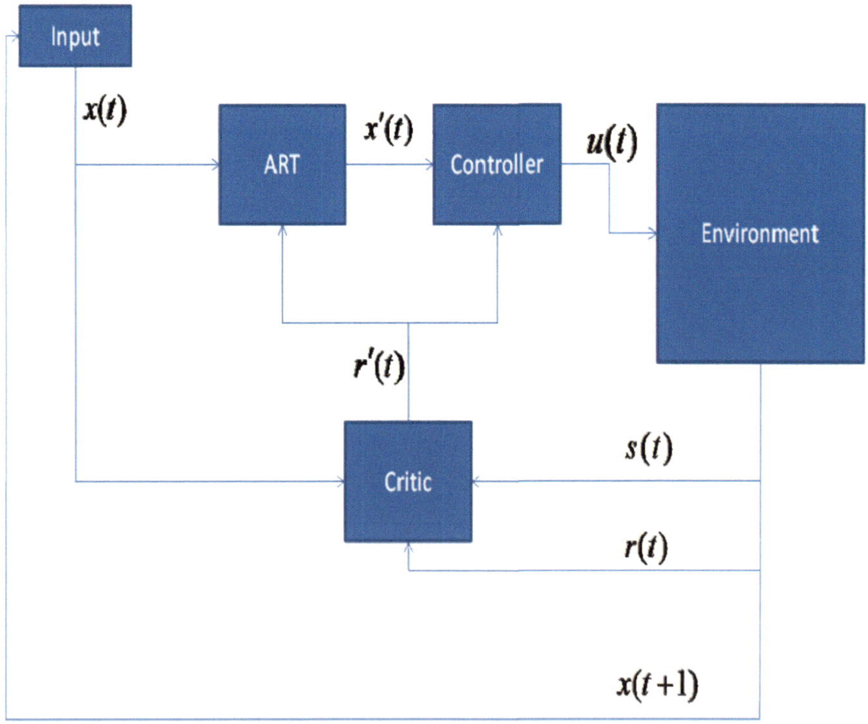

Fig. 2.1 Unified ART Block Diagram. Block Diagram of the ART Critic Algorithm. In this diagram, x(t) is the state, x'(t) the state trace, u(t) the control, r(t) the reward, r'(t) the reward trace, and s(t) the learning signal

The signals are explained as follows. The input vector is denoted as $x(t)$, while $x'(t)$ is the input trace. This architecture contains two sets of adaptive weights: The weights W_j drive the ART block, and the weights V_j are in the Controller. (This is different from other neural-network based Adaptive Critic designs where the value function is the sole purview of the Critic unit.)

Details of the steps in the block diagram are as follows:

Step 1: Calculate State Trace

a) Calculate category choice vector $T_j(x(t))$.
b) Run vigilance test on T_j.
c) Set state trace $x'(t)$ equal to the winning T_j node.

Step 2: Calculate Control

a) Select action label for j.
b) Check V_j for label and take a weighted sum of T_j and V_j.
c) Set the control $u(t)$ equal to the chosen label.

Step 3: Process Control

a) Submit $u(t)$ to the environment.
b) Observe $x(t+1)$, $r(t)$, and $s(t)$.

Step 4: Interpret Reward via Critic

Case 1: If supervisory signal is present, then use it to update the weights W_j

a) if $s(t) = u(t)$, then set $V_{j,s(t)}$ to the max value $\bar{V}$, zero out all the other values of V_j, and update the weights $W_j = W_j \wedge x(t)$.

b) if $s(t) \neq u(t)$, trigger reset in ART. Get a new $x^{'}(t)$ and repeat step (a). If $x^{'}(t)$ is an uncommitted node, then set it as in (a).

Case 2: If reinforcement signal is positive, update V_j using ADP methods and update $W_j = W_j \wedge x(t)$.

Case 3: If reinforcement signal is negative, trigger mismatch in ART and select new $x^{'}(t)$ in a manner proportional to $r(t)$. Then update $V_j$as in 2 and W_j as standard.

Case 4: If neither $s(t)$ nor $r(t)$ are present, then update $W_j = W_j \wedge x(t)$.

The details of these operational steps follow in the next subsection.

2.4 Operation

This section provides details for the steps outlined in the previous section.

2.4.1 Step 1: Calculate State Trace

The first step is to present the input to the ART system. Details are shown in Figure 2.2.

The category choice vector $T_j(x(t))$ is calculated following the normal ART equations given in Figure 1.1. This category choice vector will now select the category with which to match the input, and the appropriate vigilance test will be run. If the input passes the vigilance test and resonance occurs, then the winning node will be selected and set equal to the state trace signal $x'(t)$. If resonance is not achieved, then a new category will be tested, ranked in descending order in the category choice vector. If none of the categories are able to pass the vigilance test, then an uncommitted node will be recruited and assigned to the state trace. Note that when an uncommitted node is recruited, the dimensions of the controller unit must be updated along with the ART unit. Section 2.4.2 contains more details regarding the controller activation.

Alternatively, the state trace may encode a distributed activation level corresponding to each category. In this case, the category choice vector $T_j(x(t))$ combines with the results of the vigilance test to generate a state trace $x'(t)$, which is a vector with a value for each encoded node. Such a distributed input may achieve better performance in applications with noisy inputs if category proliferation becomes an issue.

The state trace is the transformed measure of input. The trace will key the proper activation levels in the controller and will be susceptible to modification by the critic in response to a reinforcement signal. It is here, in the ability of the supervised learning mode and especially of the reinforcement learning mode to access the resonance stage of the ART architecture, that the unified ART model presented herein truly remains a member of the biologically inspired computational family.

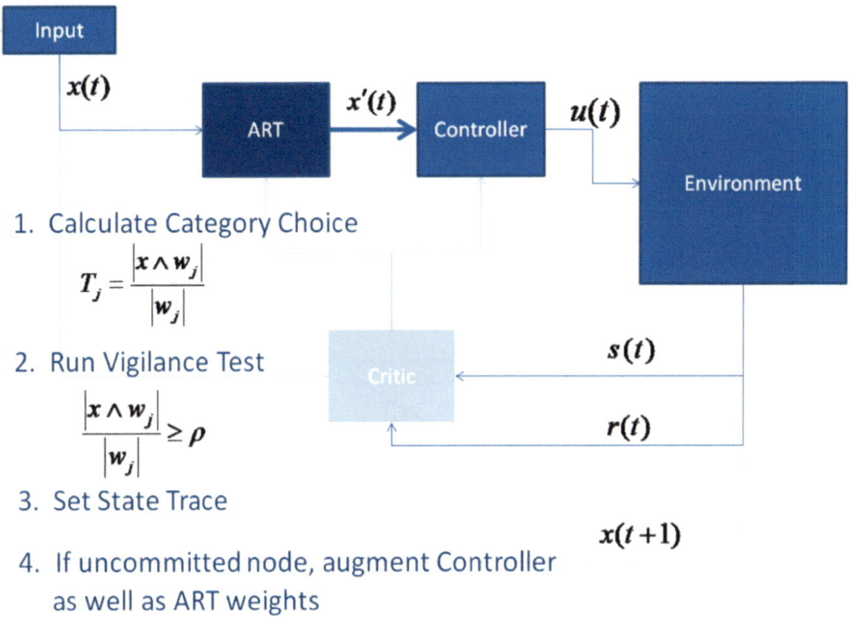

Fig. 2.2 The State Trace Calculation Step of the Unified ART Architecture

2.4.2 Step 2: Calculate Control

The controller, as shown in Figure 2.3, is a matrix V_{jk}, where the columns represent categories and the rows the various control (or action) signals.

The controller, having a number of rows equal to the committed nodes in the ART unit, must be updated whenever the state trace $x'(t)$ is set to an uncommitted neuron. In this structure, the controls represent the possible inputs to the environment unit. In much of the reinforcement learning literature, particularly that surrounding the use of actor-critic networks, the controls are interpreted as *actions* an agent may take at any state in the environment. In keeping with the more general control theoretic framework, we adopt the term *control* even though it is within an actor-critic model.

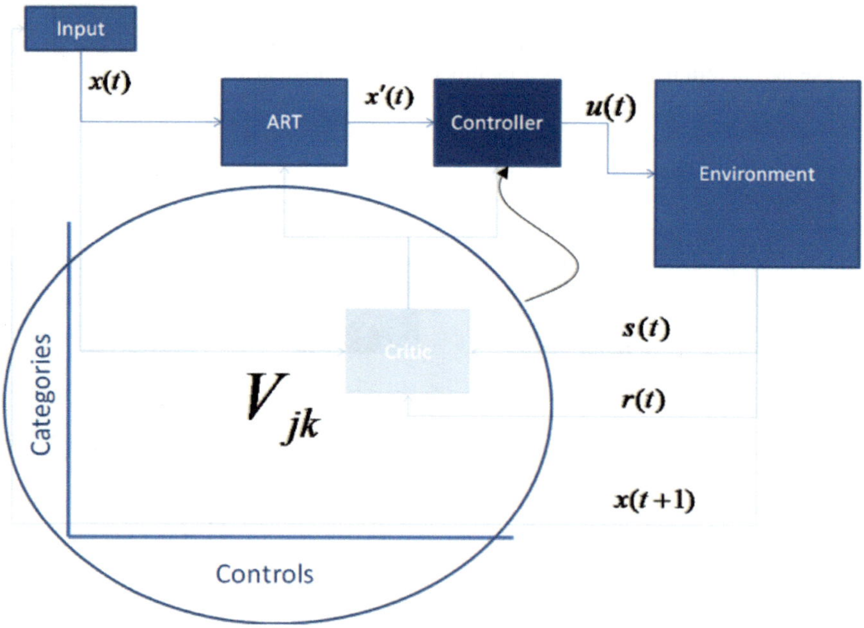

Fig. 2.3 The Controller in the Unified ART Architecture

In order to calculate which control will be applied to the environment, the state trace is used to select a row in the control matrix. The entries of the controller serve as a value function telling us which of the available controls the system believes best applies in the given state. The choice is the entry in the selected row with the greatest value. This entry is then assigned as the control signal $\mu(t)$ and submitted to the environment.

If the alternative distributed encoding of the state trace $x'(t)$ is being used, then the control is calculated in a different way. Figure 2.4 shows the operation of the second step using the alternative extended version of the state trace vector.

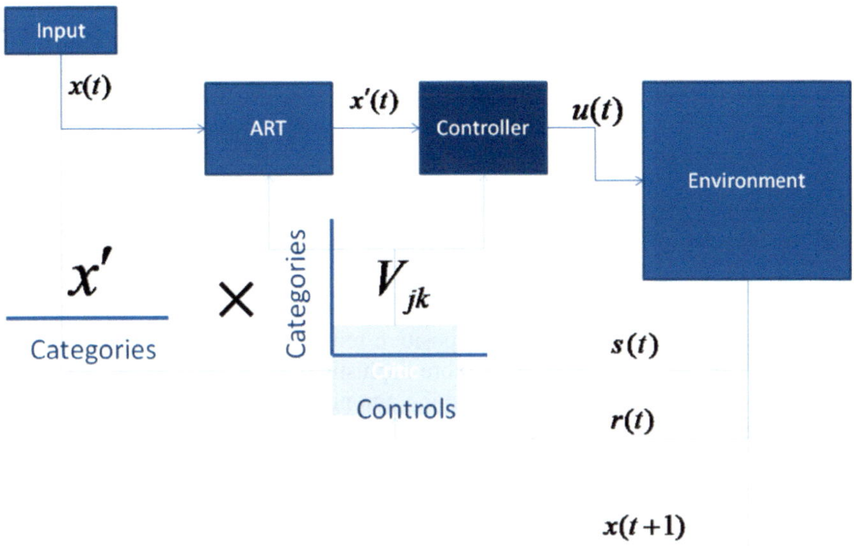

Fig. 2.4 The Distributed State Trace

For a distributed state trace, the control selection requires a matrix multiplication, with the appropriate row in the controller being effectively weighted by the values in the processed category choice vector. The resulting weights are then inspected for the highest value, as is common in value function representation schemes, and the winner chosen as the control signal $\mu(t)$.

Either method of calculating the control will result in the generation of a signal which is then presented to the environment.

2.4.3 *Step 3: Process Control*

The environment responds to the presentation of a control signal by generating three signals: the supervisory signal $\boldsymbol{s(t)}$, the reinforcement signal $\boldsymbol{r(t)}$, and the next state $\boldsymbol{x(t+1)}$. The supervisory signal has the capacity to override the state trace calculated through the resonance process in Step 1. Similarly, a negative reinforcement signal is also given that authority.

These signals can be seen in the system block diagram in Figure 2.1. How each of them is handled is covered in the next subsection.

2.4.4 Step 4: Interpret Reward via Critic

This step involves four cases: (1) supervisory signal, (2) positive reinforcement, (3) negative reinforcement, (4) unsupervised mode.

2.4.4.1 Supervisory Signal

If the environment produces a supervisory signal, then this overrides any other signals present. While an environment should not produce both supervisory and reinforcement signals, in the event this does occur, the unified ART system will give priority to $\boldsymbol{s(t)}$, as it represents an input-output pairing that simply must be attended to. In comparison, even a positive reinforcement signal does not carry such a mandate. (Note that while more sophisticated incarnations of the unified ART system are capable of distinguishing among and rating multiple supervisors, the base version treats the supervisory signal as infallible.)

If $s(t) = u(t)$, then the system has generated the correct control. In this case, the value of this control $V_{j,s(t)}$ is set to the maximum allowable value $\bar{V}$; all other values of V_j are zeroed out. The adaptive ART weights are updated according to the standard rule $W_j = W_j \wedge x(t)$.

If the supervisor signal does not match the system's chosen output, that is, if $s(t) \neq u(t)$, then a reset is triggered in the ART unit. During reset, the category choice vector T_j has its highest value zeroed out, and the vigilance test resumes with the next highest node. After the vigilance test selects another winning node, the state trace value $x'(t)$ is reset, and the algorithm returns to Step 2.

2.4.4.2 Positive Reinforcement

If $\boldsymbol{r(t)} > 0$, then the ART system has returned a good, but perhaps not perfect, control signal. The controller weights $\boldsymbol{V_j}$ are updated by adding the reinforcement signal (or a value proportional to $\boldsymbol{r(t)}$) to the entry corresponding to the chosen control. This is the standard actor-critic framework update process. Compared to the case in which the supervisory signal is present, the value of the control is not automatically set to the maximum possible while simultaneously zeroing out all others, thus ensuring they are not chosen. However, the already vibrant connection between the current input and the positively reinforced control signal is strengthened.

Finally, the ART weights are updated according to $W_j = W_j \wedge x(t)$.

2.4.4.3 Negative Reinforcement

In the presence of negative reinforcement, the algorithm, in order to be truly integrated, must force a reset in ART. In this way, a negative reinforcement signal from the environment interacts directly with the resonance loop. ART mismatch is handled as in Case 1, and a new state trace $\boldsymbol{x}'(\boldsymbol{t})$ is generated. The controller weights $\boldsymbol{V}_{\boldsymbol{j}}$ are updated as in Case 2, and the ART weights are standard in all cases.

2.4.4.4 Unsupervised Mode

In the absence of supervisory or reinforcement signals, the system operates in unsupervised mode. The ART weights are updated as $\boldsymbol{W}_{\boldsymbol{j}} = \boldsymbol{W}_{\boldsymbol{j}} \wedge \boldsymbol{x}(\boldsymbol{t})$, and the system proceeds to the next input. This mode is identical to pure, unsupervised ART.

2.5 An Extended Architecture

The unified learning architecture presented thus far is the one implemented in the extensive application in Chapter 3. This section introduces an extended version of the algorithm, which explicitly calls upon the canonical features of the ART family of systems to showcase more directly the theoretical developments of Section 1.5.

2.5.1 The Vigilance Test

The basic structure of an ART algorithm takes the following canonical form:

1. Calculate coding node activity T_j.
2. Run vigilance test by optimizing T_j subject to the match criterion $\mathcal{V}_i$.
3. Update adaptive weights.

The vigilance test, with its interplay between bottom-up activity and top-down expectation, defines an ART algorithm. Different specializations of ART have different forms for T_j and $\mathcal{V}_i$, and while they may also specify the optimization procedure to use when running the vigilance test, they all contain this inherent search for a resonant state. Furthermore, vigilance tests can be layered for more complicated architectures, as is the case when combining learning modes. In these situations, multiple match criteria must be met, each one representing a layer of the test. For this reason, the match criteria are expressed as a vector.

In the ART algorithm discussed in this chapter, $\mathcal{V}_i$ is allowed to depend on the learning mode while still adhering to the canonical structure.

For all three learning modes, define

$$T_j = |x \wedge W_j| + (1 - \alpha)(M - |W_j|) \tag{2.1}$$

where x is the input (state), α is the signal rule parameter (typically $\propto = .01$), M is the size of the input vector, $|\cdot|$ denotes the L_1-norm, $\wedge$ is the fuzzy intersection operator defined by $(p \wedge q)_i = \min(p_i, q_i)$, W is the coding node weight matrix, and j indexes the categories.

The vigilance tests vary by learning mode.

For unsupervised learning,

$$\mathcal{V}: |x \wedge W_j| \geq \rho|x| \tag{2.2}$$

where ρ is the vigilance parameter. No special rules are imposed on the vigilance test. Typically, the values of T_j are just sorted and then checked against $\mathcal{V}$ in descending order until the criterion is met. Some researchers report computational gains by first checking the elements of T_j against $\mathcal{V}$ and then sorting, but the result will be the same either way.

If none of the nodes pass the vigilance test, then a new node is committed. See Section 2.5.2 for details on this procedure.

If the system is running only in unsupervised learning mode, then its operation defaults to Fuzzy ART.

For supervised learning, use the following vigilance test:

$$\begin{aligned} \mathcal{V}_1&: |x \wedge W_j| \geq \rho|x| \\ \mathcal{V}_2&: u = \mathcal{S} \end{aligned} \tag{2.3}$$

where u is the control output and $\mathcal{S}$ is the supervisory signal. For this vigilance test, not only is it the goal to satisfy the condition given in the UL vigilance test, but also to ensure that the output is correct. This is an example of layered vigilance testing as introduced in ARTMAP.

To perform this test, a procedure called *match-tracking* is specified. During match-tracking, modify the vigilance according to

$$\rho = \frac{|x \wedge W_j|}{M} + \varepsilon \tag{2.4}$$

and keep searching T_j until a node passes the criteria. In this updated equation, ε is a small decrement which defaults to $\varepsilon = -.001$. The vigilance parameter is returned to its baseline value at the conclusion of the vigilance test.

If none of the nodes pass the vigilance test, then a new node is committed as outlined in Section 2.5.2. If our system runs only in supervised learning mode, its operation roughly emulates that of Default ARTMAP.

For reinforcement learning, calculate the temporal difference (TD) error given by

$$\delta = \mathcal{R} + \gamma V(x_{t+1}) - V(x_t) \tag{2.5}$$

where $\mathcal{R}$ is a reinforcement signal, γ is a learning rate, x_t and x_{t+1} are the current and next states, respectively, and V is the value weight matrix.

When the TD-error is high, the system wants to increase the chance that the given u will be selected. When the TD-error is low, it wants to decrease that chance that the given u will be selected. The second layer of the vigilance test reflects this design, while the first layer represents the core ART vigilance test. The ideal match criteria are given by

$$\begin{aligned} \mathcal{V}_1 &: |x \wedge W_j| \geq \rho|x| \\ \mathcal{V}_2 &: \delta > 0 \end{aligned} \tag{2.6}$$

Alternatively, one could increase expectations and require δ greater than some high reward value. In any case, the optimization method for this $\mathcal{V}_2$ is actually a *suboptimization* procedure. In fact, the value of δ never changes during the vigilance test; therefore, if $\mathcal{V}_2$ is not satisfied on the first pass, it will never be satisfied. In order to properly optimize $argmax(T_j)$ subject to (2.06), the entire T_j vector must be searched thoroughly as is done in the vigilance tests for unsupervised and supervised learning. However, in the reinforcement learning case, more than just the current state is needed; the next state is also required. The only way to get the next state is to submit u to the environment repeatedly until the control that gives the greatest value for δ is discovered. Therefore, the vigilance test cannot require searching the entire T_j vector in a manner that will allow the continual calculation of new δ's. Rather, it can only search T_j to select a new u and then use that information as best as the system can.

So, for the case in which $\delta \geq 0$, the system refrains from searching T_j at all and simply updates the weights. If $\delta < 0$, however, then implement the following plan:

1. Trigger a reset in ART by zeroing out the current winning node and selecting the next highest node that passes $\mathcal{V}_1$.
2. Calculate a new u.
3. If the new u is different than the old u, then update weights using the new winning category.
4. If the new u is the same as the old u, then trigger reset in ART and search again.
5. If all the categories have been searched and u remains unchanged, then commit an uncommitted node and update weights using the newly committed category as the winning node ω.

Updating U and V in the case of negative reinforcement will reduce the chance of the system again choosing u when the given category wins. However, it does not increase the chance that the given state input will be placed in the same category the next time it is seen.

2.5.2 The Weight Update

Fuzzy ART uses one set of weights to perform unsupervised learning. ARTMAP adds a second layer of weights to implement supervised learning. Our algorithm continues the trend and adds a third weight matrix to handle reinforcement learning. Each set of weights has its own role to play and its own update rules.

The coding node weights W determine the templates for the ART categories. W is a $j \times k$ matrix, where j indexes the number of categories and k indexes the size of the state vector.

The update rule for W is as follows:

$$W_\omega = \beta(x \wedge W_\omega) + (1 - \beta)W_\omega \tag{2.7}$$

where ω is the winning category node and β is the coding node learning parameter. When $\beta = 1$, the updates are called *fast learning*. This is its usual and default value.

The next set of weights to discuss are the control weights U. Default ARTMAP calls these weights the output layer weights, but in this application the outputs correspond to controls, hence the need for the renaming. U is a $j \times c$ matrix, where j indexes the categories and c the controls. The elements of U represent values for state-control pairs. The control is selected by choosing the column with the largest value, and U can be considered an *actor*, in the language of ADP. This matrix is affected heavily by the supervised learning mode.

The update rules for U are as follows:

1. If $\mathcal{S}$ is active, then

$$U_{\omega,:} = U_{\omega,:} y^{\mathcal{S}} \tag{2.8}$$

2. If $\mathcal{R}$ is active, then

$$U_{\omega u} = \min\left(U_{\omega u} + \lambda\mathcal{R}, \bar{u}\right) \tag{2.9}$$

where $U_{\omega,:}$ is the ωth row vector in U, the *control mask* $y^{\mathcal{S}}$ is the row vector whose elements i are zero when $i \neq \mathcal{S}$ and equal to the maximum control value $\bar{u}$ divided by $U_{\omega i}$ when $i = \mathcal{S}$, u is the control output, and λ is the reinforcement learning rate. Note that the result of the notationally challenging expression $U_{\omega,:} = U_{\omega,:}y^{\mathcal{S}}$ is simply to set the values in the winning node's weight vector: the correct control, the one that matches the supervisory signal $\mathcal{S}$, is set to the maximum allowable value $\bar{u}$, and the other signals are set to zero.

The control weights U are not updated during unsupervised operation.

The final set of weights used in the extended ART Critic architecture is the value weight matrix V. The value weights act as a *critic* for the actor U. These weights determine the TD-error δ, which is used to update both U and V. These weights, like U, store state-control values. The difference is that U is heavily influenced by the supervisory signal $\mathcal{S}$ while V is only affected by the reinforcement signal $\mathcal{R}$. The value weights are of the same dimension as the control weights.

The update rule for V is as follows:

$$V_{\omega} = V_{\omega} + \lambda\mathcal{R} \tag{2.10}$$

ART algorithms only commit new memory to the storage of template and other weight values when it becomes necessary to do so. If a vigilance test fails for all the categories, then a new category, previously uncommitted, is committed and has the current input assigned to it as a template. Since all the weight matrices W, U, and V have a row for each category, committing a new node requires altering their structure. For W, the new row is initialized to all $\mathbf{1}$'s, and for U and V, the new row is initialized to small random values. The variable K tracks the number of categories, so it must be incremented as well.

Finally, the nature of the environment must be addressed, tying into the view of unified learning presented in Section 1.5. The learning problem under consideration requires the agent to interact with its environment. The environment is modeled by a vector-valued function of the form

$$f(x_t, u) = (x_{t+1}, \mathcal{S}, \mathcal{R}) \tag{2.11}$$

For a pure unsupervised or supervised learning problem, thinking of an "environment" may not seem as natural as simply presenting to the algorithm a sequence of input vectors. In this case, the x_{t+1} output of f can be considered a virtual state equal to the next input on the list. In the case of a more traditional environment, the next state is the natural evolution of the system under the control u as determined by internal dynamics.

An unsupervised learning problem will not contain either $\mathcal{S}$ or $\mathcal{R}$ signals. In a supervised problem, the $\mathcal{S}$ signal is equal to the target class, and for reinforcement learning, the $\mathcal{R}$ signal is the traditional reinforcement value. Note that there is a difference between these signals not being active and these signals being zero.

2.5.3 Algorithm

To summarize the operation of the extended unified ART architecture, the algorithm is presented in the following convenient list:

1. Calculate coding node activity using (2.01).
2. Select winning category $\omega = argmax\ (T_j)$.
3. Choose control signal $u = \max\ (U_\omega)$.
4. Submit u to the environment and receive x_{t+1}, $\mathcal{S}$, and $\boldsymbol{\mathcal{R}}$.
5. If neither $\mathcal{S}$ nor $\boldsymbol{\mathcal{R}}$ is active, run the unsupervised vigilance test (2.02).
6. If $\mathcal{S}$ is active, run the supervised vigilance test (2.03).
7. If $\boldsymbol{\mathcal{R}}$ is active, run the reinforcement vigilance test (2.06).
8. Update weights as given by (2.07), (2.08), (2.09), and (2.10).
9. Repeat from Step 1 with the new state.

Chapter 3
An Application of Unified Computational Intelligence

3.1 Overview

The previous section described a unified computational intelligence learning architecture based on Adaptive Resonance Theory neural networks. In this chapter, this architecture is used in an application that was briefly introduced in Chapter 1.

The content of this chapter is adapted from a paper appearing in the *Neural Networks* journal (Brannan, Seiffertt, Draelos, & Wunsch, 2009) and a preliminary version appearing as (Brannan, Conrad, Draelos, Seiffertt, & Wunsch, 2006).

In this chapter, the unified computational intelligence algorithm is referred to as CARTMAP, for Coordinated ARTMAP. This name was determined by Sandia National Laboratories collaborators since they, being government officials, possess a certain *je ne sais quoi* for the use of acronyms. This application, in the area of situation awareness, proved a testing bed for a unified learning architecture of the type described in Chapter 2 of this book. The results indicate that the task described, if performed using only a single mode of learning, would not have achieved the same level of effectiveness as it did using all three modes in combination.

3.2 Introduction

Modern information sources to support decisions in domains such as force protection are diverse. Ground, air, and space-based sensors continue to increase in capability. Information fusion algorithms can help integrate a variety of sensor data into meaningful forms (Hall & Llamas, 1997). Applications with a complex assortment of data continue to challenge machine learning approaches to information fusion, which normally utilize a single type of learning algorithm and therefore limit the use of all available data (Brannon, Conrad, Draelos, Seiffertt, & Wunsch, 2006). Our approach coordinates multiple learning mechanisms to accommodate environments where ground-truth and feedback may not be available consistently, and it uses Adaptive Resonance Theory (ART)-based networks, which are based on understanding cognition. This ties the work into

J. Seiffertt & D.C. Wunsch: Unified Computational Intell. for Complex Sys., ALO 6, pp. 33–48.
springerlink.com

other such computational architectures seeking not only solutions to engineering problems but also an understanding of the function of the brain and mind as discussed by Werbos, Perlovsky, and others.

3.2.1 Machine Learning

Machine learning involves programming computers to optimize a performance criterion using example data or past experience (Alpaydin, 2004). Artificial neural networks are commonly used in machine learning and utilize supervised, unsupervised, and reinforcement learning approaches to achieve predictive properties based on example (training) data. Unsupervised learning (clustering) can be effective when ground truth is not available within a dataset. Supervised learning (learning with a teacher) provides a means of using experience (examples with ground-truth) to correctly classify yet unseen situations. Reinforcement learning offers promise for machine learning in difficult learning environments by taking advantage of feedback about a system's performance. The challenge addressed by the current work is to coordinate all of these learning mechanisms and utilize the appropriate one based only on available information, not human intervention.

Neural networks offer an excellent assortment of high-performance, low-cost, distributed processing options. In particular, they can be embedded into appropriate sensors for operation at the lowest levels of information fusion with effective but low-complexity designs. At the highest levels of information fusion and situation assessment, reinforcement learning can be used with a human in the loop to provide operational feedback. Dealing with multiple sensor modalities and extracting meaningful information from massive datasets is a natural fit for these adaptive methods. Although neural networks have been applied to sensor fusion, their use in situation awareness has been limited, possibly because of the lack of rich training data for this problem.

Automated (computational) information fusion continues to suffer from very specific, ad-hoc solutions (i.e., there appears to be very little general-purpose technology to apply to this problem) (Kokar, Tomasik, & Weyman, 2004). For many applications, there is also a dearth of data to use for training a computational engine. This reveals a challenge for the application of machine learning techniques, which are data-driven and require training, whether via supervised, unsupervised, or reinforcement learning. On the other hand, because they are data-driven, the advantage of machine learning techniques is that they can learn solutions to problems that are difficult for humans to codify with explicit rules or models. In other words, they can represent rules/decisions that are implicit in the training data.

3.2.2 Information Fusion

The fusion of information has been likened to the ability of animals to utilize multiple senses to derive a better understanding of a situation (Hall & Llinas, 1997). For example, one may hear a noise and, based on the sound pressure discrepancy between each ear, localize the area of the sound source. Vision can then be used to further define and understand the source of the sound. The analogy is helpful because fusion, and more generally situation assessment, is a process rather than simply a discrete event. The process leads one from raw data to understanding and actionable knowledge. Fusion can occur over various information (sensor) modalities, over geographic space, and over time.

The sources of information potentially available to decision makers continue to expand in depth and breadth. Sensor capabilities in particular are maturing rapidly, but a valid concern is that the pace of sensor development has not necessarily been consistent with advances in human effectiveness, which the sensors must ultimately support (Paul, 2001). Fusion algorithms will better support human-in-the-loop system effectiveness when the decision maker is a central and balanced design element. Our system includes, as a core design principle, the use of a human-in-the-loop operator to provide reinforcement signals as well as to ensure a level of quality control.

3.3 Approach

3.3.1 System Architecture

The design of the computational engine for information fusion and situation awareness takes advantage of the diverse utility of neural networks and integrates elements of supervised, unsupervised, and reinforcement learning. The design not only advances machine learning research, but also addresses the needs of situation awareness and human-in-the-loop decision support.

Key design attributes of our system include accepting various inputs such as binary, categorical, and real-valued data. With respect to situation assessment outputs, attributes include confidence levels as well as evidence in support of or against the assessment. In the context of missing or noisy inputs, the system exhibits graceful performance degradation.

In order to address the desired design attributes of our situation awareness system, neural networks are employed for information fusion, followed by a situation assessment module. ARTMAP is based on Adaptive Resonance Theory (ART), a widely implemented approach to modeling the learning capabilities of the brain (Carpenter & Grossberg, 1988). Architectures based on ART have been used successfully in a variety of areas requiring a self-organizing pattern recognition neural network. The basic ART element supports unsupervised learning and binary inputs. Fuzzy ART is an extension to accommodate categorical and real-valued inputs. ARTMAP supports supervised learning and can accommodate real-valued inputs using fuzzy logic (Carpenter, Grossberg, Markuzon, Reynolds, & Rosen, 1992). ARTMAP can also support reinforcement learning, for example, by adding a mechanism to implement actor-critic methods. Coordinated ARTMAP (CARTMAP) is the name given to the current approach and involves the integration of all three learning mechanisms in the same architecture.

The situation assessment module receives state information from the information fusion module and possibly other sources and outputs a threat assessment or action to be taken.

3.3.2 Information Fusion Engine

Intelligent creatures exhibit an ability to switch seamlessly among unsupervised, supervised, and reinforcement learning as needed. However, machine learning architectures, including artificial neural networks, have not yet achieved this goal. The current research contends that it is advantageous to develop this capability in a computational framework and that the ART architecture is an excellent choice for such an implementation.

A well designed sensor fusion algorithm, like an intelligent creature, can make informed use of all three types of learning on the data set given. Certain information fusion paths may be pre-trained prior to deployment, thus granting the human operators license to verify that the most obvious sensor patterns will be classified successfully. During operation, a reinforcement signal provided either by the environment or by the human operator acting off of the fusion algorithm's recommendations can adjust the current adaptive weight profile to curtail or retrain a faulty clustering (negative reinforcement) or to promote successful clustering (positive reinforcement) in the ART algorithm. Finally, in the absence of any external signal, the algorithm will learn in an unsupervised manner, comparing current inputs to what it already knows.

With the ARTMAP unit taking the place of the actor in the actor-critic implementation, the Coordinated ARTMAP (CARTMAP) algorithm behaves according to the following steps:

1. Upon receipt of an unsupervised signal, the system uses its exemplar classification scheme (the ART unit) to output an action choice, as usual. No updating of the lookup table will be necessary.
2. When presented with a supervised signal, the internal adaptive weights update as per normal ARTMAP rules, and the output action is set equal to the supervised training signal. Furthermore, the values in the lookup table for actions not associated with the supervisory signal are zeroed out.
3. When a reinforcement learning input signal is received, it will be interpreted according to the Q-learning algorithm. The appropriate entry in the lookup table is augmented with the new reinforcement value, and the action selected is the one with the most value accumulated in its column of the table. In the simulations, the values of the parameters delta and gamma are 0 and 1, respectively.

In summary, the information fusion engine accepts raw data from sensors and other information sources and processes/transforms/fuses them into inputs appropriate for the Situation Awareness Assessment engine.

The information fusion system utilizes appropriate elements of its architecture based on the data presented to it. The three ART networks are linked together by an inter-ART module (Associative Memory). One ART unit handles the inputs, another ART unit processes the supervisory (or target) signal, and the other processes the reinforcement signal as an adaptive critic. This architecture is capable of online learning without degrading previous input-target relationships.

There are times when unsupervised learning is satisfactory, such as in the presentation of new input vectors to a pre-trained network. Supervised learning is appropriate and desired for initial training on fixed data. However, these two types of learning do not cover every possible complication. There are times when the human operator does not know the correct classification, yet some feedback on the decision can be provided. These situations fall into the reinforcement learning category. One aspect of developing this information fusion engine, therefore, is adding the reinforcement learning capability to the ARTMAP neural network.

3.4 Application

The situation awareness system was designed to operate in an environment involving distributed sensors and a central collection site for protection of a facility. Information sources in such an environment can include seismic, magnetic, acoustic, passive infrared (PIR), and imaging sensors as well as weather, time/day information, various intelligence information, local/regional/federal threat levels or law enforcement bulletins, and any other information that might be relevant to the security of a particular facility, such as current traffic situations or health issues.

Conditions of interest to force protection decision makers include: no activity, severe weather, unauthorized people or vehicles in certain locations, and certain types of unauthorized vehicles or humans with weapons in any areas. Actions include: doing nothing, identifying the type and location of a moving object (vehicle or human), using commands to turn sensors on or off, dispatching forces, and/or notifying higher authorities. The information sources can include binary data, such as motion detection, categorical data, such as the type of day (weekend, holiday, etc.), and real-valued time-series data, such as seismic, acoustic, and magnetic energy levels.

Before being deployed, the system must be pre-trained with any information the human operator knows about the system. For example, if the data signature of a thunderstorm is easy to demonstrate (due to specific acoustic, magnetic, etc. levels), then that information can be included in the supervised training portion of the system. The information fusion engine will adaptively learn many more data-observation relationships during online operation, but having basic readings pre-trained will aid in initial operation.

When an intruder, be it an unauthorized vehicle or a human with a weapon, breaches the sensor range of a protected facility, the triggered sensor data stream into the information fusion engine. The CARTMAP network then maps these data into observations, such as a vehicle heading north at high speed. These pairings represent novel data readings that were not anticipated, which are then categorized via the CARTMAP algorithm in relation to the pre-trained data.

The observation is then sent to the situation assessment engine, which follows the partially observable Markov decision process (POMDP) formulation to calculate a probability distribution over the state space. This information represents a confidence level that the system is in any given state. The state with the highest confidence from this calculation represents the system's choice for the current state. All of this probability information is then passed to the human operator, who uses this evidence in making a final decision about how to respond to the situation.

Adapting online is an important element of the system and is accomplished through reinforcement signals that can be sent through the system in two ways. First, if the probabilities of each state are so low that the human operator would not be able to distinguish the state from simple background noise, then the situation assessment engine may issue a command to gather more information from additional sensors. Second, the human operator may disagree with the system's assessment of the current state. A reinforcement signal is then sent to the information fusion engine, and the data-observation mappings will adapt online. Both of these reinforcement signal loops are noted functionally in the block diagram in Figure 3.1. This feature of the system allows it to maintain relevance in a changing environment.

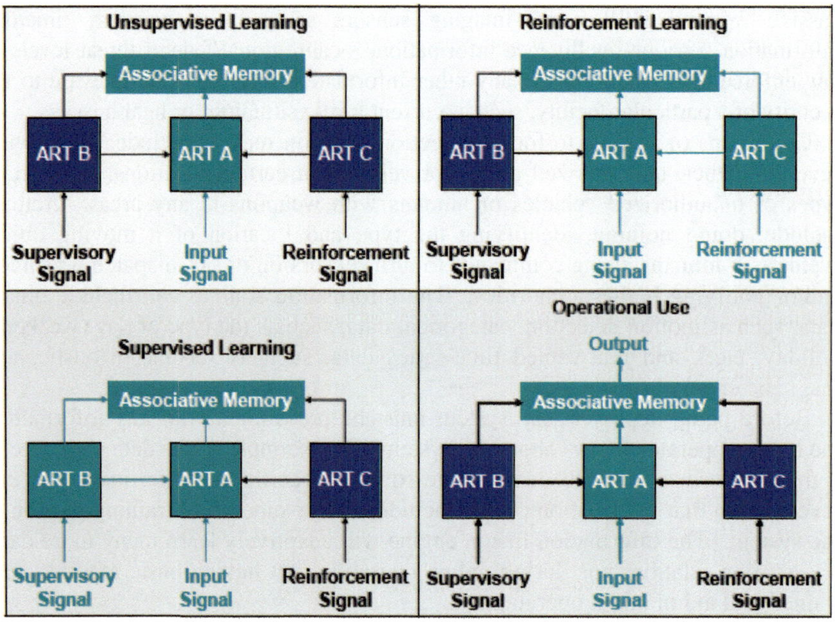

Fig. 3.1 CARTMAP Input and System Activity. Associated with unsupervised learning, supervised learning, reinforcement learning, and standard operational use. Available inputs to the system are shown in green, as are the active elements involved in learning

As shown in Figure 3.1, unsupervised learning occurs using a single ART unit. The cluster that forms is the one that maximizes the signal strength of the input with respect to a match criterion. Many forms of both the signal and the match criterion are used in various implementations of an ART architecture. Amis and Carpenter (2007) provide default values that work in general scenarios. Supervised learning occurs when the clusters formed by the unsupervised learning unit are given labels through interaction with supervisory inputs. This interaction is mediated by an associative learning field as explained in Amis and Carpenter (2007). This process forces a reset in the input cluster if the label does not match the supervisory signal closely enough. Finally, reinforcement learning is handled in a similar manner. The RL signal can update the associate weights following the Q-learning explained in Section 3.3.2.

The CARTMAP algorithm was implemented in Matlab and applied to information fusion in a vehicle tracking scenario that is described in more detail below. ART is at the core of the fusion engine. During off-line training, an input pattern is presented to the CARTMAP network and, depending on its similarity to existing category templates, it is either assigned to a current winning category or a new category is created for it. Categories may exist indefinitely without an assigned class. However, if a supervisory signal accompanies the input, the target class is immediately associated with the category. During offline reinforcement learning, an input pattern is presented to the CARTMAP network, and a winning category is determined. A reinforcement signal is computed based on the class of the winning category and the ground-truth class. For example, if the category's class matches the ground-truth class, the reinforcement signal is assigned a positive reward; if not, then a penalty is assigned. A range of reinforcement values are assigned based on the quality of the match. A reinforcement lookup table (RLUT) is used to track an input pattern's relationship with possible classes. The RLUT stores input patterns and an accumulated reinforcement signal for each possible class. CARTMAP weights are updated according to the following criteria:

1. If no category encodes the input pattern, then a new category is created without a class assignment.
2. If the winning category has an unassigned class, then the RLUT is searched for the input pattern. If the pattern is found in the RLUT, then the reinforcement signal is applied to the class of the winning category, and the class with the highest reinforcement is used as the target in supervised learning. If the pattern is not found in the RLUT, then nothing is done to the CARTMAP weights.
3. If the winning category has an assigned class, then this class and reinforcement signal are used by a critic function to determine how to update CARTMAP weights. The RLUT is searched for the input pattern. If the pattern is not found, unsupervised learning is performed, and the pattern is added to the RLUT along with the reinforcement signal. If the pattern is found in the RLUT, then the reinforcement signal is applied to the RLUT for the class of the winning category, and the class with the highest reinforcement is used as the target in supervised learning.

The decision support graphical user interface (GUI) consists of three screens. The center screen is primarily imagery (i.e., from cameras, photography augmented with graphics, and/or fully synthetic renderings) (see Figure 3.2). The second screen displays a log of temporal track data (see Figure 3.3). The log reflects temporal features, such as how long ago an unauthorized vehicle breached a sensor field and how soon another track might reach a key threshold (e.g., a fence or different sensor field). The third and most detailed screen provides track detail and assessment bases (see Figure 3.2).

The log screen and track detail screen utilize features found in the Tactical Decision Making Under Stress (TADMUS) system (Morrison, Kelly, Moore, & Hutchins, 1997). The TADMUS system has motivations similar to the current research in that more content needs to be devoted to supporting an understanding of a given context. In both TADMUS and our situation awareness approach, less emphasis is placed upon evaluating possible courses of action.

The track detail GUI provides typical track parameters such as an object's course and speed, but significant detail is provided with respect to the basis for assessment. Evidence in support of and against a given assessment is displayed. The machine learning algorithms share the evidence used to derive assessments with the operator. Such an approach provides greater transparency and allows the operator to interrogate assessments.

For the example scenario of an unauthorized vehicle, the assessment could be a "threat." Evidence in support of such an assessment includes sensor data such as explosives detected, but also local law enforcement data such as the license plate returning as a stolen vehicle. Evidence against the assessment could include a relatively slow speed and the use of the vehicle for construction when there has been ongoing construction activity. Alternative assessments are shown along with their respective evidence in support of or against them.

The operator can investigate various assessments along with corresponding courses of action. For example, a patrol vehicle in the vicinity of the unauthorized vehicle could be directed closer to the possible threat. Further, other types of sensors can be activated to generate additional points of reference and work towards higher levels of assessments, such as possible intent.

3.4.1 Vehicle Tracking

The situation awareness technology was applied to tracking vehicles in the vicinity of a facility under force protection. A data set suitable for testing and demonstrating our technology was collected during a DARPA SensIT program in November, 2001 at Twenty-Nine Palms, CA and exists at the University of Wisconsin (UW) (Duarte & Hu, 2004). The data set consists of raw time series (acoustic and seismic) and binary detection decisions from 23 sensor nodes distributed along three intersecting roads as one of two vehicles travels along a road. Figure 3.2 includes a map illustrating the force protection scenario, with a fence line and an Entry Control Point (ECP) providing protection for a facility on the North Road. The two vehicles used in the scenario are a light armored vehicle (AAV) and a heavier, tracked transport vehicle (DW). A scenario was developed whereby a facility under protection is assumed to exist along one of the roads, and binary sensor data processed by our fusion and situation assessment algorithms are used to inform a human decision maker.

Fig. 3.2 Vehicle Tracking Scenario Map. Blue dots are seismic/acoustic sensor nodes. The speed, heading, location, and vehicle type are estimated by independent CARTMAP networks using binary data from all sensor nodes as input

3.4.2 Analysis

This section provides analysis of the experimental results.

3.4.2.1 Force Protection Experiments

In order to demonstrate the capabilities of the situation awareness system, neural networks were trained to perform sensor fusion, a situation assessment formula was constructed/calculated, and a GUI was developed, all to increase the awareness of a human decision maker of the situation around the facility under their protection. The scenario consists of a virtual checkpoint partway up the north road on the way to a sensitive facility with 23 sensor nodes scattered along three intersecting roads. Each sensor node outputs a binary detection decision at fixed time intervals (0.75 seconds in the original test set). The sensor detections derive from seismic, acoustic, and passive infrared energy levels. The (AAV and DW) vehicles move from one end of a road, through the intersection, and to the end of another road. The total number of runs is 40, which includes 20 original data sets from the SensIT experiment. An additional twenty runs were created by artificially reversing the direction of the vehicle. This is possible by simply presenting the data in reverse. In other words, the sensor record from the last time step would be presented to the information fusion system first, the first time step would be presented last, and so on for all the time steps in the run. It is plausible that the information is accurately represented in these runs because the data consists of binary decisions and the ground is relatively flat, so the engine speed and noise are presumably similar in both directions.

The primary piece of information that a decision maker wants to know is the current threat level around his facility. The threat level is a function of the location, speed, heading, and type of vehicle detected by the sensor array and other variables that are independent of the sensor array, such as Department of Homeland Security (DHS) advisory level, wind speed, average batter level of the sensors, time of day, and day of week.

The system used to produce the threat level is illustrated in Figure 3.3. The system consists of three modules: 1) Information Fusion, 2) Situation Assessment, and 3) a Graphical User Interface (GUI) focused on human decision makers in force protection applications. Multiple time steps of binary sensor data serve as input to the Information Fusion module, which implements the CARTMAP algorithm. This introduces an element of relative time, which is a necessary component in estimating speed and heading. The output from the Fusion module consists of vehicle type, speed, location, and heading, each with a corresponding confidence level, and will serve as input to the Situation Assessment module. This module consists of rules that represent the conditions under which a threat is defined. The output of the assessment module will feed the graphical user interface (GUI) with a threat level (low, medium, high), an associated confidence level, a suggested response, and evidence in support of or against its output. The GUI will also have access to the output from the fusion module, maps, and other available data, such as time, date, and environmental data. All elements of the situation awareness system were implemented in Matlab and tested with the vehicle tracking data from UW in the force protection scenario just described.

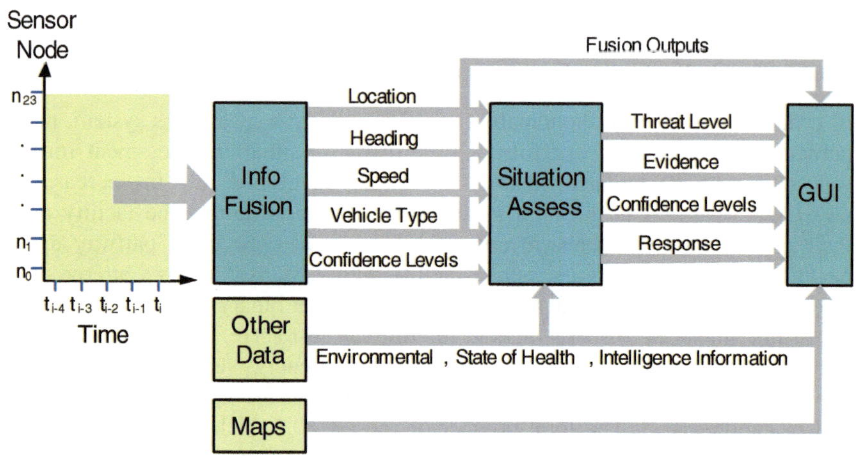

Fig. 3.3 Force Protection Experiment Using UW Vehicle Data. Multiple time steps of binary sensor data are used as input to the CARTMAP Information Fusion module. Vehicle information from the Fusion module and other additional data are used as input to the Situation Assessment module, which outputs actionable information to the user

3.4.2.2 Results of Training the Fusion Model

The fusion model consists of four different CARTMAP networks, one for each fusion output (location, heading, speed, and vehicle type). The output of a network will be of a categorical type or class except for the confidence levels, which will be real numbers. Table 3.1 presents the classes for each information fusion network. Note that for each network, if the input is all zeros, the output will be zero by virtue of a simple fixed rule (i.e., no learning is involved).

Out of the 40 total runs available for the force protection experiments, 70% were used for training and the remainder for testing. Table 3.2 shows the number of runs used in the six experiments. In real-world applications, it is expected that the amount of supervised training data is limited. In the force protection experiments, only 2 of the 28 training runs are used for supervised learning.

Table 3.1 Information Fusion Output Classes for the Four CARTMAP Networks. (Vehicle Type, Location, Heading, and Speed)

Vehicle Type Classes	Location Classes	Heading Classes	Speed Classes
0: zero input	0: zero input	0: zero input	0: zero input
1: AAV	1: West Road	11: N	1: < 10 km/hr
2: DW	2: North Road	14: NE	2: 10-20 km/hr
	3: East Road	13: E	3: 20-30 km/hr
	4: Intersection	8: SE	4: 30-40 km/hr
		4: S	5: 40-50 km/hr
		1: SW	6: 50-60 km/hr
		2: W	7: 60-70 km/hr
		7: NW	8: 70-80 km/hr
			9: 80-90 km/hr
			10: > 90 km/hr

Table 3.2 Distribution of Vehicle Runs Used to Experiment with Different Learning Modes. Experiments 1/2/3 and 4/5/6 use the same data, but use learning modes in a different order

Experiment #	# Supervised Runs	# Unsupervised Runs	# Reinforcement Runs	# Test Runs
1 & 4	2	26	0	12
2 & 5	2	13	13	12
3 & 6	2	0	26	12

Experiments 1-3 use the same runs as Experiments 4-6, but the order of training is reversed. In Experiments 1-3, supervised learning is conducted first, followed by reinforcement learning, and finally unsupervised learning. Experiments 4-6 use the opposite order of learning, using the data with the least amount of information first and finishing with supervised learning, which utilizes training data with the most information. In this case, one expects the richer data sets and training modes to correct errors and refine the classification performance of previous learning modes.

For each force protection experiment conducted, the same test set was used, consisting of 12 runs with 1755 input/output pairs. The performance (% correct classification) was computed based on this test set. For some sensor modes, such as speed and heading, a classification error may not necessarily indicate poor performance. For example, if the ground truth heading of a vehicle is North and the fusion module output is Northeast, it would be counted as a classification error even though the output is quite satisfactory. Experiments 1-6 were conducted using various combinations of learning modes for each of the information fusion networks. The best results for each network are presented in Table 3.3.

In the Classified Correct (%) column of the tables, there are three numbers separated by colons (e.g., 1 : 2 : 3). The numbers in position one represent the percentage of test samples that have a target value exactly matching the output value from a CARTMAP network.

The numbers in the second position represent the percentage of test samples that have a target value exactly or partially matching the output value from a CARTMAP network. An exact match increments the total number of correct classifications by 1, whereas a partial match increases the number by 0.5. Partial matches are possible only with the Heading and Speed networks, where the class adjacent to the target class is considered a partial match. For example, if the target class is N, then a network output of NW or NE would result in a partial match. Note that for the Vehicle Type and Location networks, no partial matches exist, so the first and second numbers in the Classified Correct column should be the same.

Table 3.3 The Best Fusion Test Results of the Four CARTMAP Networks. Reinforcement learning followed by supervised learning worked best for estimating vehicle type and location, while supervised learning followed by unsupervised learning and then reinforcement learning worked best for vehicle heading and speed. In the Classified Correct (%) column of the table, there are three numbers separated by colons (e.g., 1 : 2 : 3). The numbers in position one represent the percentage of test samples that have a target value exactly matching the output value from a CARTMAP network

Sensor Mode	Experiment #	Learning Mode	Vigilance	# Categories	Classified Correct (%)
Vehicle Type	6	Reinforcement Supervised	0.7 0.65	36 : 108 44 : 112	92.6 : 92.6 : 91.7 92.7 : 92.7 : 91.7
Vehicle Location	6	Reinforcement Supervised	0.7 0.65	22 : 58 31 : 61	96.8 : 96.8 : 98.0 96.9 : 96.9 : 98.0
Vehicle Heading	2	Supervised Reinforcement Unsupervised	0.9 0.7 0.7	39 45 : 59 45 : 59	68.4 : 69.6 : 69.6 66.6 : 79.9 : 80.3 62.7 : 75.8 : 81.7
Vehicle Speed	2	Supervised Reinforcement Unsupervised	0.9 0.7 0.7	46 53 : 77 53 : 77	72.4 : 79.9 : 79.9 74.3 : 82.1 : 82.0 73.4 : 81.3 : 81.8

The numbers in the third position represent correct classification percentages of networks that have had two passes through the training set. During the first pass, the reinforcement lookup table is updated during reinforcement learning. The updated table may be an advantage for second pass unsupervised and reinforcement learning. Correct classification percentages are computed using partial matches. Each network was trained using vigilance parameters that resulted in a reasonable number of categories.

In the next section, a weighted rule for determining the threat level of the situation awareness system is discussed. The rule combines the outputs of the fusion module and environmental conditions, and its output is categorized into High, Moderate or Low threat based on human judgment. Ground truth exists for the threat level, so performance of trained fusion networks with specified environmental conditions can be measured. Two environmental conditions are specified: 1) Benign – each environmental condition is set to its lowest value, and 2) Severe – each environmental condition is set to its highest value. For each of the learning modes, the correct classification percentage is measured against ground truth. The results are given in Table 3.4.

In practice, if only unlabeled data is available, then machine learning is typically not used at all. Machine learning is most often used when some labeled data are available and supervised learning is then used to its maximum extent, while other learning techniques are not employed. The advantage of using a variety of machine learning techniques is evident in Table 3.3 and Table 3.4 above, but a single set of networks (possibly a different network for each sensor mode) must be chosen since one cannot generally anticipate the environmental conditions. Table 3.5 summarizes the performance results of using the best combination of supervised (SL), unsupervised (UL), and reinforcement learning (RL) in comparison to the more common use of supervised learning alone. Table 3.6 lists the machine learning approaches used by each CARTMAP network to produce the best situation assessment threat level performance averaged over benign and severe environmental conditions.

Table 3.4 Best Test Results of Situation Assessment Threat Level Performance. Using a combination of learning modes under benign and severe environmental conditions. Different learning modes for different CARTMAP fusion networks are necessary to produce the best situation assessment results

Environment Condition	Vehicle Exp #	Location Exp #	Heading Exp #	Speed Exp #	Reinforcement Iterations	Classified Correct (%)
Benign	1	2	3	3	1	88.9
Benign	1	3	2	3	2	89.5
Severe	1	2	2	6	1	86.8
Severe	3	2	5	3	2	87.7

An important conclusion drawn from the experimental results is that utilizing multiple training approaches that can take advantage of additional and different data produces superior results for situation awareness compared to supervised training alone. The reason performance decreases with UL after SL is that with SL alone, all test patterns get encoded by a labeled category, whereas after UL, there are now unlabeled categories that may encode test patterns producing classification errors. Even though these unlabeled categories sometimes lowered the performance, they may eventually add value after subsequent labeling during SL or RL. Unsupervised input patterns that get encoded by existing categories with a class label can contribute to the quality of the category in representing the class in feature space. In addition, since the CARTMAP has access to a reinforcement lookup table (RLUT), if an unlabeled pattern matches a pattern in the RLUT, the corresponding class label from the RLUT can be assigned to the unlabeled pattern. This feature is used during unsupervised learning. Originally, the RLUT is generated from the supervised training data. It expands when new unlabeled patterns are encoded by categories with class labels and the pattern and its label are added to the RLUT.

Table 3.5 CARTMAP Fusion Performance Results. Using multiple machine learning modes in comparison to supervised learning alone

Learning Approach	Vehicle %	Location %	Heading %	Speed %	Avg. Threat %
SL	81.8	95.6	69.6	79.9	78.5
SL with UL and/or RL	92.7	98.0	81.7	81.9	87.6

Table 3.6 The Combination of Learning Approaches. The combinations that produced the best threat level performance. Three different combinations were used for the four different fusion modules (vehicle type, location, heading, and speed)

	Vehicle	Location	Heading	Speed
Learning Approach	SL, UL	SL, RL	SL, UL, RL	SL, RL
Reinforcement Iterations	1	2	2	2

Results for Experiment 3 (SL followed by RL) reveal a strong relationship between the hints that RL provides and partial matching in scoring the classification performance. When exact classification matches are required, hints may not be good enough. However, if a "close enough" match is sufficient, then RL hints improve performance. Even though multiple vigilance values were used in the force protection experiments, it is expected that performance will improve when the vigilance is optimized for the type of fusion mode and the type of learning.

It is important in RL to have data representing all classes that a network is designed to classify. If a class is not represented in the data, RL will not be able to establish a label for this class.

Vehicle location is the easiest piece of information to learn with binary sensor data. Location is inherent in the sensors themselves because their position is fixed.

Since 54.4% of the input patterns are all zeros, if a correct classification percentage of greater than 54.4% is achieved after UL only, then the reinforcement lookup table is being used to correctly label some patterns. During reinforcement learning, an input pattern is submitted to a network, and a reinforcement signal is generated. This signal offers negative or positive feedback on the output of the network. The following steps are taken at this point of reinforcement learning. When a reinforcement signal is received, the RLUT is updated, and SL is performed if the input pattern is found in the RLUT (the action associated with the input pattern with the highest value is used as the target). Unsupervised learning is performed if the input pattern is not found in the RLUT and the reinforcement signal is positive.

In general, SL should be used to create as many categories as possible within reason, while subsequent non-supervised training should take advantage of these existing categories and enrich them without corrupting them. The coordination of three machine learning modes therefore offers potential benefit from every sample of data available in an application.

3.5 Future Work

Arguably the most immediate area of future work lies in establishing principles and practices for employing the three learning modes. There are different ways of combining three modes of machine learning and many options for how and when to employ each mode. The current research offers a preliminary perspective on leveraging each learning mode for greatest system performance. It stands to reason that a CARTMAP network can be tailored for each information fusion mode (vehicle type, speed, heading, and location). The vigilance parameter may be different for each mode and may also require adjustment based on the type and ordering of the learning modes.

The core of our machine learning approach is an ART neural network. Other algorithms and architectures should be explored with the same goal in mind, that of integrating multiple learning modes. Reinforcement learning is a general area of research worth pursuing in the area of situation awareness where there is often not a clear win or lose outcome by which to measure success. There are also many ways of performing reinforcement learning, some closer to supervised learning, with stronger hints, and others that provide rare but consistent hints about the system's performance. How many iterations to use in reinforcement learning on this problem is a legitimate research question, as is how best to acquire feedback from human decision makers or the overall force protection system, either directly or indirectly.

Another avenue of future machine learning research is to explore the use of ensembles or bagging for supervised learning (Dietterich, 2000). The use of ensembles employs multiple "experts" that train the same network using a different sampling with replacement from the original supervised training set. The combination of the experts' solutions results in higher performance than the use of a single network trained on the original data set.

3.6 Conclusion

The coordination of the three major machine learning approaches in a single architecture, using ARTMAP at its core, is an innovation that should prove valuable in addressing real-world problems. Many domains offer a limited amount of information with ground truth that can be used with supervised learning algorithms. More available is data with hints from the environment that can be used with reinforcement learning. Almost always, data is available without labels that can be used with unsupervised learning. Allowing these three modes of learning to be used in the same framework is an important contribution. Interesting advantages emerge when these three approaches leverage one another. For example, reinforcement learning can utilize supervised learning when enough information about class labels is available from the environment. Unsupervised learning can take advantage of stored reinforcement learning information to go beyond mere clustering. There is potential for interplay between the learning modes that does not exist with a single mode.

Chapter 4
The Time Scales Calculus

4.1 Introduction

This chapter begins the second part of this book. The first part outlined a unified computational intelligence learning architecture based on neural networks. The design, theoretical underpinnings, and an application were presented to achieve the first goal of this book: to develop unified computational intelligence for *learning*.

This chapter focuses on another of the goals: to develop unified computational intelligence for *adapting*. Now, the difference between learning and adapting may be negligible to some and semantics to others; certainly, the potential for overlap exists, and it is not necessary to develop a categorical definition such that any given algorithm can be classified as learning, adapting, or neither. For example, in order for the unified ART architecture to learn new sensor signatures, it must in a sense adapt to its environment. But in this chapter, what is meant by *adapt* is more general and divorced from learning in the sense of neural networks as function approximators. *Adapting* refers to a system that makes decisions in a complex and unpredictable environment. It is this decision making that is of the most interest, and the field to be discussed now is one that grew out of the need to optimize solutions to multi-stage decision processes. It is called dynamic programming, and although a cousin to operations research, it is firmly a member of the computational intelligence family.

This part of the book will discuss decision making under the dynamic programming framework in some detail. The discussion here is deeply mathematical and technical.

What is being unified here is the nature of the time scale under which decisions are made—under which agents can *adapt*. Traditional computational decision theory is broken into discrete-time and continuous-time models. To unify these approaches, advantage is taken of a new and rapidly developing field of mathematics called the time scales calculus. Within this field, researchers seek to draw together analysis of both discrete and continuous intervals. They search for what unifies differential and difference equations and operators and look to the construction of dynamical systems completely independent of the nature of the

J. Seiffertt & D.C. Wunsch: Unified Computational Intell. for Complex Sys., ALO 6, pp. 49–60.
springerlink.com

underlying signals. Still in its infancy, this mathematics has much room for growth. Indeed, at times the work in this book develops new mathematical tools in order to derive the dynamic programming results. However, the promise for applications is high, as researchers continue to take on the control of systems that combine discrete and continuous signals.

This chapter provides background on the essentials of the time scales calculus, which will be used to derive the results. This material is adapted from (Seiffertt & Wunsch, 2007), (Seiffertt, Sanyal, & Wunsch, 2008a), (Seiffertt, Sanyal, & Wunsch, 2008b), (Seiffertt & Wunsch, 2008), and (Seiffertt, 2009). The dynamic programming results are presented in Chapter 5, and the unification of neural network learning, necessary to supplement the dynamic programming approaches, is presented in Chapter 6.

4.2 Fundamentals

Traditional analysis and discrete mathematics work with functions defined on a domain that is either entirely discrete or continuous. The study of the time scales calculus allows for the consideration of functions on domains that can be a mixture of the two. As such, *dynamic equations* in a general sense are discussed rather than specifying difference or differential equations. This section presents enough of the theory of dynamic equations to support the results in this book; this coverage includes definitions of characteristic functions, dynamic derivatives of single and multiple variables, and basic integration on time scales. In addition to the established mathematics, this book provides new contributions to this theory: the definition of σ_n-complete differentiability and a chain rule for n-variables. These results are used in Chapter 6 to prove results for neural network learning.

Dynamic equations are defined on mathematical structures called *time scales*. Formally, a time scale (T,α) is an ordered pair such that $\mathbb{T} \subset \mathbb{R}$ is nonempty, every Cauchy sequence in $\mathbb{T}$ converges to either a point within $\mathbb{T}$ or to a finite infimum or supremum of $\mathbb{T}$, and α is a function from $\mathbb{T}$ into $\mathbb{T}$. This function α, along with the nature of the time scale under consideration, will dictate the properties of the chosen dynamic derivative and the resulting dynamic equations.

Sets qualifying as time scales include the integers $\mathbb{Z}$, the natural numbers $\mathbb{N}$, the scaled integers $h\mathbb{Z}$ (for $h > 0$), the quantum calculus time scale $q^{\mathbb{N}_0}$ for $q > 1$ (the quantum calculus, a specialization of the time scale calculus, will be considered at length), limiting sets such as $\{0\} \cup \left\{\frac{1}{n} : n = 1,2,...\right\}$, finite unions of intervals such as $[0,1] \cup [2,3] \cup [5,6]$, and unions of intervals and discrete points such as $\{1,2,3,4,5\} \cup [5,8]$. More exotic sets, such as the Cantor set, are also technically time scales, although they may arouse less interest from application-minded researchers. Furthermore, well-known sets such as the rationals $\mathbb{Q}$and the irrationals $\mathbb{R}\backslash\mathbb{Q}$, and open intervals such as $(0,1)$, are *not* time scales.

Four characteristic functions are defined on time scales, two that scroll through the time scale and two that detail the degree of continuity of the time scale. The *forwards* and *backwards jump operators* $\sigma(t)$ and $\rho(t)$ give us the "next" and "previous" elements in a time scale, respectively. They are defined as follows:

$$\begin{aligned} \sigma(t) &= \inf\{x \in \mathbb{T}: x > t\} \\ \rho(t) &= \sup\{x \in \mathbb{T}: x < t\} \end{aligned} \tag{4.1}$$

For $\mathbb{T} = \mathbb{R}$, $\sigma(t) = \rho(t) = t$, and for $\mathbb{T} = \mathbb{Z}$, $\sigma(t) = t + 1$ and $\rho(t) = t - 1$. More complicated time scales admit more involved jump operators. Consider, for example, the q-time scale studied in the quantum calculus (Kac & Cheung, 2002): $\mathbb{T} = q^{\mathbb{N}_0}$ (where $q > 1$). The forward and backward jump operators are given by $\sigma(t) = q^{t+1}$ and $\rho(t) = q^{t-1}$, respectively. Additionally, it is often convenient to write $f(\sigma(t))$ as $f^{\sigma}(t)$ and $f(\rho(t))$ as $f^{\rho}(t)$.

These jump operators are a critical determinant of the character of a dynamic equation. For example, to work with the commonly used delta derivative, consider $\mathbb{T}$ to be any nonempty closed subset of the real line, and set α equal to the forward jump operator σ. Similarly, when working with the nabla derivative time scales of the form $(\mathbb{T}, \rho)$, where ρ is the backwards jump operator, are studied. However, there is not a technical restriction to such domains in the general case, and indeed, it is this freedom that gives the study of alpha derivatives, where α is neither a forward nor a backward jump, its conceptual power and showcases the versatility of the generalized time scales approach to analysis. When the foundations of dynamic programming on these generalized time scales are established in Chapter 5, decision problems will be able to be modeled in ways that go beyond that of classical restrictions to intervals or integers.

The remaining two characteristic functions defined on time scales are referred to as the *forward and backwards* step functions μ and ν. These functions trace the level of continuity of a time scale and are defined as follows:

$$\begin{aligned} \mu(t) &= \sigma(t) - t \\ \nu(t) &= \mathrm{t} - \rho(t) \end{aligned} \tag{4.2}$$

For continuous domains, the step functions are $\mu(t) = \nu(t) = 0$, and for the integers they are $\mu(t) = \nu(t) = 1$. For the q-time scale, $\mu(t) = q^{t}(q - 1)$. The forward step function $\mu(t)$ is often called the *graininess function* in the literature, usually when the research at hand is limited to the forward, or delta, dynamic derivative.

It is important, since both discrete and continuous intervals are permitted in the domain, to speak of a property of the points of a time scale called *density*. In the traditional calculus, a point is considered *dense* if there is, in some sense, an infinity of points close to it. Every point in the real line is dense, and none of the integers are dense. A point that is not dense is said to be isolated or scattered. On dynamic domains, there may be some dense points and some isolated points. Furthermore, it is possible to encounter transition points, which are labeled as *right-* or *left- dense* or *scattered.* Formally, these concepts are defined as follows for a point $t \in \mathbb{T}$:

- t is *right-dense* if $t < \sup(\mathbb{T})$ and $\sigma(t) = t$
- t is *left-dense* if $t > \inf(\mathbb{T})$ and $\rho(t) = t$
- t is *dense* if t is both right-dense and left-dense
- t is *right-scattered* if $t < \sigma(t)$
- t is *left-scattered* if $t > \rho(t)$
- t is *isolated* if t is both right-scattered and left-scattered

Since dynamic equations are defined on more complex domains than their differential and difference equation counterparts, it is important to keep track of the nature, either dense or scattered, of each point.

From a time scale $\mathbb{T}$, derive three sets $\mathbb{T}^\kappa$, $\mathbb{T}_\kappa$, and $\mathbb{T}^\kappa_\kappa$ as follows: $\mathbb{T}^\kappa = \mathbb{T}\backslash\{a\}$ if $\mathbb{T}$ is bounded above and a is left-scattered, $\mathbb{T}_\kappa = \mathbb{T}\backslash\{b\}$ if $\mathbb{T}$ is bounded below and b is right-scattered, and $\mathbb{T}^\kappa_\kappa = \mathbb{T}^\kappa \cap \mathbb{T}_\kappa$. These sets form the domains of the dynamic derivatives discussed in the next section.

4.3 Single-Variable Calculus

The usual derivative of the time scales calculus is defined as follows. Let $f: \mathbb{T} \to \mathbb{R}$ be a function. Then the delta derivative $f^\Delta(t)$ of f at a point $t \in \mathbb{T}^\kappa$, where $\mathbb{T}^\kappa$ coincides with $\mathbb{T}$ except at a left-scattered maximum, if one should exist, is defined to be the number such that given $\varepsilon > 0$ there is a neighborhood U of t such that

$$|[f(\sigma(t)) - f(s)] - f^\Delta(t)[\sigma(t) - s]| \leq \varepsilon|\sigma(t) - s| \tag{4.3}$$

for all $s \in U$, where *neighborhood* is defined such that $U = (t - \delta, t + \delta)$ for some $\delta > 0$. Note that this follows the classical definition of the derivative, with the traditional $x + h$ increment replaced by the forward jump operator $\sigma(t)$. This sort of translation is common in the calculus of time scales. The delta derivative $f^{\Delta}(t)$ becomes $f^{'}(t)$ when $\mathbb{T} = \mathbb{R}$ and becomes the standard difference operator on $\mathbb{T} = \mathbb{Z}$.

There is also a backwards derivative, defined as follows. Let $f: \mathbb{T} \to \mathbb{R}$ be a function. Then the nabla derivative $f^{\nabla}(t)$ of f at a point t is given by the number, provided it exists, such that given $\varepsilon > 0$ there exists a neighborhood U of t such that

$$|[f(\rho(t)) - f(s)] - (\rho(t) - s)f^{\nabla}(t)[\sigma(t) - s]| \leq \varepsilon|\rho(t) - s| \tag{4.4}$$

for every $s \in U$. At left-scattered points, this becomes the left difference operator found in the traditional study of difference equations.

When $\mathbb{T} = \mathbb{R}$, both the delta and nabla derivatives reduce to the traditional derivative $f'(t)$. The terms $\sigma(t) - s$ and $\rho(t) - s$ take the place of the $x - h$ construction in the classical calculus. To get a feel for the delta and nabla derivatives, consider the following facts:

- If f is left-continuous at a right-scattered point t, then f is delta differentiable at t
- If f is right-continuous at a left-scattered point t, then f is nabla differentiable at t
- If t is right-dense, then f is delta differentiable at t if the limit $\lim_{s \to t} \frac{f(t)-f(s)}{t-s}$ exists
- If t is left-dense, then f is nabla differentiable at t if the limit $\lim_{s \to t} \frac{f(t)-f(s)}{t-s}$

It is also possible to define a more general alpha derivative. Let $(\mathbb{T}, \alpha)$ be a generalized time scale and let $f: \mathbb{T} \to \mathbb{R}$ be a function. Then the alpha derivative $f^{\alpha}(t)$ of f at a point t is given by the number, provided it exists, such that given $\varepsilon > 0$ there exists a neighborhood U of t such that

$$|[f(\alpha(t)) - f(s)] - (\alpha(t) - s)f^{\alpha}(t)[\alpha(t) - s]| \leq \varepsilon|\alpha(t) - s| \tag{4.5}$$

for every $s \in U$.

For all of these dynamic derivatives, many of the standard rules for derivatives pertain. To illustrate this, the anticipated sum, product, and quotient rules for alpha derivatives are shown:

$$\begin{aligned}(f+g)^{\alpha}(t) &= f^{\alpha}(t)+g^{\alpha}(t)\\ (fg)^{\alpha}(t) &= f^{\alpha}(t)g(t)+f^{\alpha}(t)g(t)\\ \left(\frac{f}{g}\right)^{\alpha}(t) &= \frac{f^{\alpha}(t)g(t)-f(t)g^{\alpha}(t)}{g(t)g^{\alpha}(t)}\end{aligned} \tag{4.6}$$

One crucial result from classical analysis, however, does not hold for dynamic derivatives: the chain rule. Instead of a single formula to handle differentiation of compositions of functions, dynamic equations rely on a suite of different rules, each dependent on the character of the functions. The most general of these for functions of a single variable, and the one that will be used in the formulation of the Hamilton-Jacobi-Bellman equation and the backpropagation update equations, is due to Potzsche (2002) and is stated as follows: Let $f: \mathbb{R} \to \mathbb{R}$ be continuously differentiable and let $g: \mathbb{T} \to \mathbb{R}$ be delta differentiable. Then $f(g(t))$ is delta differentiable, with $(f \circ g)^{\Delta}(t)$ given by the following integral equation:

$$(f \circ g)^{\Delta}(t) = g^{\Delta}(t)\int_0^1 f'(g(t)+h\mu(t)g^{\Delta}(t))dh. \tag{4.7}$$

When $\mathbb{T} = \mathbb{R}$, then $\mu(t) = 0$, and the equation reduces to our expected chain rule. Thus, the forward step function (or graininess) of the dynamic domain under consideration has a great effect on the emergent calculus. It turns out that many results of standard analysis, including the Hamilton-Jacobi-Bellman equation and backpropagation equations, are dependent on the use of the traditional chain rule.

A similar result holds for the nabla derivative. With $g: \mathbb{T} \to \mathbb{R}$ nabla differentiable,

$$(f \circ g)^{\nabla}(t) = g^{\nabla}(t)\int_0^1 f'(g(t)+h\mu(t)g^{\nabla}(t))dh. \tag{4.8}$$

In addition to this chain rule, which will be used during derivations, this book also provides two new chain rules, one for the multivariate case and one for the ordered derivative case.

Integration of dynamic derivatives must be considered during the derivation of the Hamilton-Jacobi-Bellman equation. A function $f: \mathbb{T} \to \mathbb{R}$ is *right-dense continuous*, or *rd-continuous*, if it is continuous at right-dense points and its left-sided limits exist at left-dense points. Similarly, a function is *left-dense continuous*, or *ld-continuous*, if it is continuous at left-dense points and its right-sided limits exist at right-dense points. The following two fundamental results hold:

- If f is rd-continuous, then

$$\int_a^t f(\tau)\Delta\tau = F(t) - F(a) \tag{4.9}$$

- If g is ld-continuous, then

$$\int_a^t g(\tau)\nabla\tau = G(t) - G(a) \tag{4.10}$$

where $F^{\Delta}(t) = f(t)$ and $G^{\nabla}(t) = g(t)$ are delta and nabla antiderivatives of f and g, respectively.

The theory of integration on time scales is not delved into deeply here. Instead, the presentation of these formulas will suffice for our results. For a thorough overview of integration theory on time scales, the reader is directed to Bohner & Guseinov, 2005, Bohner & Peterson, 2003, and Guseinov, 2003.

This concludes the introduction of the single variable time scales calculus. The next section attends to the multivariate case.

4.4 Calculus of Multiple Variables

This book uses a definition of partial derivatives on time scales given by Jackson, 2006. Let $\mathbb{T}_1, \mathbb{T}_2, \dots, \mathbb{T}_n$ be time scales, set $\mathbb{T} = \mathbb{T}_1 \times \mathbb{T}_2 \times \cdots \times \mathbb{T}_n$, and let $f: \mathbb{T} \to \mathbb{R}$ be a function. Define the operators on $\mathbb{T}$ as $\sigma(t) = (\sigma(t_1), \sigma(t_2), \dots, \sigma(t_n))$ and $\rho(t) = (\rho(t_1), \rho(t_2), \dots, \rho(t_n))$. Also define $\mathbb{T}^{\kappa} = \mathbb{T}_1^{\kappa} \times \mathbb{T}_2^{\kappa} \times \cdots \times \mathbb{T}_n^{\kappa}$, $f^{\sigma_i(t)} = f(t_1, t_2, \dots, t_{i-1}, \sigma_i(t_i), t_i, \dots, t_n)$, $f^{\rho_i(t)} = f(t_1, t_2, \dots, t_{i-1}, \rho_i(t_i), t_i, \dots, t_n)$ and $f_i^s = f(t_1, t_2, \dots, t_{i-1}, s, t_i, \dots, t_n)$.

The *partial delta derivative of f at t with respect to t_i* is the number f^{Δ_i}, provided it exists, such that given any $\varepsilon > 0$ there exists a neighborhood $U$of t_i for $\delta > 0$ such that

$$|[f^{\sigma_i}(t) - f_i^s(t)] - f^{\Delta_i}(t)[\sigma_i(t) - s]| \le \varepsilon|\sigma_i(t) - s| \tag{4.11}$$

for all $s \in U$, where *neighborhood* is defined such that $U = (t_i - \delta, t_i + \delta) \cap \mathbb{T}_i$.

In a similar way, the *partial alpha derivative of f at t with respect to t_i* is the number f^{α_i} such that given $\varepsilon > 0$ there exists a neighborhood $U$of t_i for $\delta > 0$ such that

$$|[f^{\alpha}(t) - f_i^s(t)] - f^{\alpha_i}(t)[\alpha_i(t) - s]| \le \varepsilon|\alpha_i(t) - s| \tag{4.12}$$

Higher order partials can be defined in the usual way. It is even possible to consider mixed nabla, delta, and alpha partial derivatives. Further details can be found in Bohner & Guseinov, 2004.

Necessary for the proof of the HJB equation on time scales is the chain rule for partial derivatives given by Bohner and Guseinov [18]. Let $t \in \mathbb{T}$, $x: \mathbb{T} \to \mathbb{R}$,

$y: \mathbb{T} \to \mathbb{R}$, $x(\mathbb{T}) = \mathbb{T}_x$, $y(\mathbb{T}) = \mathbb{T}_y$, and $F(x(t), y(t))$. Assume $x(\sigma(t)) = \sigma_x(x(t))$ and $y(\sigma(t)) = \sigma_y(t)$. If $F = f(x(t), y(t))$ is σ_x-completely differentiable and x and y are differentiable, then

$$F^\Delta(t) = f^{\Delta_x}(x(t), y(t))x^\Delta(t) + f^{\Delta_y}(\sigma_x(x(t)), y(t))y^\Delta(t). \tag{4.13}$$

For more details on partial derivatives on time scales, including a complete treatment of σ_x-complete differentiability, the reader is directed to Bohner & Guseinov, 2004.

Note that the development of partial dynamic derivatives is still in its infancy and is not entirely settled. Different authors use different notations for these concepts. It will be made clear in this book which conventions apply for the theoretics. The next section provides a contribution to this literature in the form of an extension of the Bohner-Guseinov chain rule to the multi-dimensional case.

4.5 Extension of the Chain Rule

An extension of this chain rule to the case of n variables will be required. As a preamble to the proof of this n-variable chain rule, the important definition of σ_1-*delta differentiability* is discussed.

For a full discussion of σ_1- *delta differentiability*, the reader is directed to Bohner and Guseinov, 2004. Here, we present one fundamental part of the definition, which will be needed in the extension of the definition to the case of n variables. In order for a function $f: \mathbb{T} \to \mathbb{R}$ to be σ_1- delta differentiable, the following condition must hold:

$$\begin{aligned} f(\sigma_1(t_1^0), \sigma_2(t_2^0)) - f(t_1, t_2) = A_1[\sigma_1(t_1^0) - t_1] + B[\sigma_2(t_2^0) - t_2] + \\ \gamma_1 [\sigma_1(t_1^0) - t_1] \end{aligned} \tag{4.14}$$

where B is the number $f^{\Delta t_2}(\sigma_1(t_1^0), t_2^0)$ and the $\gamma_i \to 0$ as $t \to t_0$. Further note that it is possible to define σ_2-delta differentiability in an analogous manner. Only required is the σ_1 version, as results obtained for that generalize immediately to the higher indexed cases.

Now, for the statement of the Bohner-Guseinov chain rule, let $t \in \mathbb{T}$, $x: \mathbb{T} \to \mathbb{R}$, $y: \mathbb{T} \to \mathbb{R}$, $x(\mathbb{T}) = \mathbb{T}_x$, $y(\mathbb{T}) = \mathbb{T}_y$, and $F(x(t), y(t))$. Assume $x(\sigma(t)) = \sigma_x(x(t))$ and $y(\sigma(t)) = \sigma_y(y(t))$. If $F = f(x(t), y(t))$ is σ_1-completely differentiable and x and y are differentiable, then

$$F^\Delta(t) = f^{\Delta_x}(x(t), y(t))x^\Delta(t) + f^{\Delta_y}(\sigma_x(x(t)), y(t))y^\Delta(t) \tag{4.15}$$

The condition $x(\sigma(t)) = \sigma_x(x(t))$ is referred to as *forward jump commutativity* and is called upon often in the chain rules for partial derivatives on time scales.

For an extension of this theorem to the case of n variables, first define the notion of σ_1-delta differentiability for n variables instead of just two. The condition (4.14) becomes

$$f(\sigma_1(t_1^0),\dots,\sigma_n(t_n^0)) - f(t_1,\dots,t_n) = A_1[\sigma_1(t_1^0) - t_1] + \sum_{i=2}^{n} B[\sigma_i(t_i^0) - t_i] + \sum_{i=1}^{n} \gamma_i\,[\sigma_1(t_1^0) - t_1] \tag{4.16}$$

Armed with this definition, the following theorem can be proven.

Theorem 4.1 (Chain Rule for Functions of n-variables)
Assume that $x_i\colon \mathbb{T} \to \mathbb{T}_i$, $x_i(t^0) = t_i^0$, $t^0 \in \mathbb{T}$, and $x_i(\sigma(t^0)) = \sigma_i(x_i(t^0))$. Let $(t^0) = f(x_1(t^0),\dots,x_n(t^0))$. Then

$$F^{\Delta}(t_0) = f_{t_1}^{\Delta}(t_1^0,\dots,t_n^0)x_1^{\Delta}(t^0) + \sum_{i=2}^{n} f_{t_i}^{\Delta}(\sigma_1(t_1^0),t_2^0,\dots,t_n^0)x_i^{\Delta}(t^0) \tag{4.17}$$

Proof
We proceed by constructing the definition of $F^{\Delta}(t^0)$ and using our hypotheses:

$$F(\sigma(t^0)) - F(t) = f(x_1(\sigma(t^0),\dots,x_n(\sigma(t^0)) - f(x_1(t),\dots,x_n(t)) \tag{4.18}$$

Now applying the forward jump commutativity condition, the right-hand side becomes

$$f(\sigma_1(x_1(t^0),\dots,\sigma_n(x_n(t^0)) - f(x_1(t),\dots,x_n(t)) \tag{4.19}$$

which, upon substitution, reduces to

$$f(\sigma_1(x_1(t^0),\dots,\sigma_n(x_n(t^0)) - f(t_1,\dots,t_n) \tag{4.20}$$

Using the definition of σ_1-differentiability with $A_1 = f_{t_i}^{\Delta}(t_1^0,\dots,t_n^0)$ and $B_i = f_{t_i}^{\Delta}(\sigma_1(t_1^0),t_2^0,\dots,t_n^0)$, gives

$$f_{t_i}^{\Delta}(t_1^0,\dots,t_n^0)(\sigma_1(t_1^0) - t) + \sum_{i=2}^{n} f_{t_i}^{\Delta}(\sigma_1(t_1^0),t_2^0,\dots,t_n^0)(\sigma_i(t_i^0) - t) + \sum_{i=1}^{n} \gamma_i[\sigma_i(t_i^0) - t] \tag{4.21}$$

Using the fact that $t_i^0 = x_i(t^0)$, we arrive at

$$\sigma_i(x_i(t_0) - t) + \sum_{i=1}^{n} \gamma_i[\sigma_i(x_i(t^0) - t]] \tag{4.22}$$

Again with the forward jump commutativity, we see that

$$x_i(\sigma_i(t_0) - t) + \sum_{i=1}^{n} \gamma_i[x_i(\sigma_i(t^0) - t]] \tag{4.23}$$

Finally, dividing by $\sigma(t^0) - t$ and taking the limit as $t_0 \to t$ yields

$$F^{\Delta}(t_0) = f_{t_1}^{\Delta}(t_1^0, \dots, t_n^0)x_1^{\Delta}(t^0) + \sum_{i=2}^{n} f_{t_i}^{\Delta}(\sigma_1(t_1^0), t_2^0, \dots, t_n^0)x_i^{\Delta}(t^0) \tag{4.24}$$

which is our desired result. ■

Note that, as a technical matter, it is important for the x_i's all to have the same domain $\mathbb{T}$ so that, in the final step, the limit can be taken on a single t. Without this restriction, the limit would have to carry across multiple t_i's, which would present a less tractable situation.

4.6 Induction on Time Scales

A form of backwards induction exists on time scales (Bohner & Peterson, 2001). Let $t_0 \in \mathbb{T}$ and $S(t)$ be a statement for each $t \in [-\infty, t_0)$ such that the following four conditions hold:

1. $S(t_0)$ is true
2. $S(t)$ being true at a left-scattered t forces $S\big(\rho(t)\big)$ to be true
3. $S(t)$ being true at a left-dense t forces $S(t')$ to be true for all t' in a left-neighborhood of t
4. $S(t')$ being true for all $t' \in (t, t_0]$ when t is right-dense forces $S(t)$ to be true

Then it can be concluded that $S(t)$ is true for all $t \in [t_0, \infty)$. There is also a forward version involving right-scattered and left-dense intervals and the forward jump operator $\sigma(t)$, but it is this backwards form that we use in the next section.

4.7 Quantum Calculus

Quantum calculus is the modern name for the investigation of the calculus without limits, which began with Euler, currently enjoys ties to abstract algebra, and has found application in the quantum mechanics literature. The book by Kac and Cheung (2002) covers many of the fundamental aspects of the quantum calculus. As this field becomes more widely researched, an increasing number of application areas are being discovered. For example, a recent study of financial derivative pricing realized a quantum calculus analog of the Black-Scholes equation (Muttel, 2007). Additionally, it has been shown that quantum calculus is a subfield of the more general mathematical field of time scales calculus.

The study of quantum calculus is concerned with a specific time scale, called the q-time scale, defined as follows:

$$\mathbb{T} = q^{\mathbb{Z}} = \{q^k : k \in \mathbb{Z}\} \tag{4.25}$$

such that $q > 1$. Dynamic equations in the quantum calculus, then, have domain $q^{\mathbb{Z}}$. It is worth noting that the quantum calculus converges to the classical continuous calculus in the limit as q approaches 1 from above.

The q-time scale analog of the forward jump operator is given by $\sigma(t) = q^{t+1}$. The graininess of the q-time scale can be shown to be $\mu(t) = q^t(q-1)$ via application of the definition of the forward jump operator σ and some algebra.

To discuss calculus on the q-time scale, a derivative needs to be defined. The *q-differential* of a function *f* is given by

$$d_q f(x) = f(qx) - f(x) \tag{4.26}$$

and the *q-derivative* of the function *f* is defined by the following expression:

$$D_q f(x) = \frac{d_q f(x)}{d_q x} = \frac{f(qx) - f(x)}{(q-1)x}. \tag{4.27}$$

Further derivatives can be defined in a manner analogous to their real counterparts. For example, the *second q-derivative* is defined as

$$D_q^2 f(x) = \frac{D_q f(qx) - D_q f(x)}{d_q x} \tag{4.28}$$

The standard rules for differentiation of products and quotients apply in quantum calculus:

$$\begin{aligned} d_q(f(x)g(x)) &= g(qx)d_q f(x) + f(x)d_q g(x) \\ &= f(qx)d_q g(x) + g(x)d_q f(x) \\ d_q \frac{f(x)}{g(x)} &= \frac{g(x)d_q f(x) - f(x)d_q g(x)}{g(qx)g(x)} \end{aligned} \tag{4.29}$$

Proving the dynamic programming algorithm will require the use of induction in the quantum calculus. Note further that for the q-time scale, only conditions 1 and 2 must be met because this time scale lacks dense points.

For the proof of ordered derivatives in the quantum calculus, differentiation with respect to a function will need to be employed. The Stieltjes integral provides a means for this in the traditional calculus, where there exists the relation $\int f dg = \int f g' dt$. A similar construction on the q-time scale is defined, where $\int f d_q g = \int f d_q g \, d_q t$. Now, as the q-time scale is a fundamentally discrete set, the integrals become expressed as summations as shown at the end of this section. This notation is presented in the more general case to maintain consistency with the relationship between the q-calculus and general time scales where the formula is given by $\int f \Delta g = \int f g^{\Delta} \Delta t$and where the symbol Δ denotes the idea of *delta differentiation*, which is the generalization of q-differentiation to any time scale. Deeper analysis of these concepts is beyond the scope of the current argument; the interested reader is directed to Bohner & Guseinov, 2005, Bohner & Peterson, 2003, and Guseinov, 2003 for further information on the theory of integration on time scales.

Let $f\colon \mathbb{T} \to \mathbb{R}$ and $g\colon \mathbb{T} \to \mathbb{R}$ be functions on a q-time scale $\mathbb{T}$, and define the following q-derivative:

$$\frac{d_q f}{d_q g} = f^{d_q g} \tag{4.30}$$

This is simply the q-time scale analog of the expression $\frac{\partial f}{\partial g} = \frac{\partial f}{\partial t} / \frac{\partial g}{\partial t}$ from classical analysis. The notation $f^{d_q g}$ will be used with ordered derivatives on time scales.

Finally, we will make one further note on the translation of integrals. Let $f\colon \mathbb{T} \to \mathbb{R}$ where $\mathbb{T} = \{t_0, t_1, \ldots\}$. Then $\int_{t_0}^{t} f(\tau)\Delta\tau = \sum_{\tau \in [t_0, t)} f(\tau)\mu(t)$, where μ is the graininess function of the time scale $\mathbb{T}$. In particular, for the q-time scale the following holds:

$$\int_{t_0}^{t} f(\tau)\Delta\tau = \sum_{i=t_0}^{i=q^{t-1}} f(q^i) q^i \, (q-1) \tag{4.31}$$

This construction is important in discussions of dynamic programming and backpropagation, as it is convenient and illuminating to first consider the general time scale formulation in some cases before delving into the particulars of the quantum calculus version.

Chapter 5
Approximate Dynamic Programming on Time Scales

5.1 Overview

In this chapter, the material overviewed in Chapter 4 is used to develop the foundations of dynamic programming on time scales. The primary results concern the derivation of the Hamilton-Jacobi-Bellman equation, the ghost in the dynamic programming machine, in this unified mathematical framework. Material in this chapter has appeared in Seiffertt & Wunsch, 2007, Seiffertt, Sanyal, & Wunsch, 2008a, Seiffertt, Sanyal, & Wunsch, 2008b, Seiffertt & Wunsch, 2008, and Seiffertt, 2009.

5.2 Introduction

The time scales calculus is an emerging key topic due to its many multidisciplinary applications. This calculus is extended to Approximate Dynamic Programming. The core backwards induction algorithm of Dynamic Programming is extended from its traditional discrete case to all isolated time scales. Hamilton-Jacobi-Bellman equations, the solution of which is the fundamental problem in the field of dynamic programming, are motivated and proven on time scales. By drawing together the calculus of time scales and the applied area of stochastic control via Approximate Dynamic Programming, two major fields of research have been connected for the first time.

The mathematics of time scales seeks to bridge the divide between the analysis of functions on discrete and continuous domains (Hilger, 1990). This calculus establishes a single unified framework for analysis of both difference equations and differential equations. Such *dynamic equations* on time scales (Bohner & Peterson, 2001), (Bohner & Peterson, 2003), have been applied in population biology (Bohner, Fan, & Zhang, 2007), quantum calculus (Bohner & Hudson, 2007), geometric analysis (Guseinov & Ozyilmaz, 2001), boundary value problems (DaCunha, Davis, & Singh, 2004), real-time communications networks (Gravagne, Marks, Davis, & DaCunha, 2004), intelligent robotic control (Gravagne, Davis, & Marks, 2005), adaptive sampling (Gravagne, Davis, DaCunha, & Marks, 2004), approximation theory (Sheng, Fadag, Henderson, & Davis, 2006), financial

J. Seiffertt & D.C. Wunsch: Unified Computational Intell. for Complex Sys., ALO 6, pp. 61–76.
springerlink.com

engineering (Sanyal, 2008), adaptive grids (Eloe, Hilger, & Sheng, 2006), switched linear circuits (Marks, Gravagne, Davis, & DaCunha, 2006), and the Kalman filter, (Wintz, 2009), among others. Due to the fact that this calculus must consider domains with all sorts of mathematical intricacies, the time scales calculus admits not one but an entire suite of dynamic derivatives. The standard derivative, the delta derivative, most closely mirrors the derivative found in traditional analysis. Other derivatives, such as the nabla and diamond-alpha, are also widely studied (Ahlbrandt, Bohner, & Ridenhour, 2000), (Bohner & Peterson, 2003), (Rogers & Sheng, 2007) and applied in various areas, particularly the study of numerical solutions of differential equations. This current work focuses on the alpha derivative, which is a more general case of the delta and nabla. As such, the theory presented herein contains the theory of the other two derivatives.

Dynamic programming (Bellman, 1957), (Bellman & Breyfus, 1962), (Bertsekas, 2001) outlines various methods for generating optimal solutions for multi-stage decision processes. The standard algorithm for dynamic programming involves a computationally intensive backwards induction update rule. In practice, many engineers working on industrial-scale applications have turned to approximation methods based on the optimal ones. Much research has been dedicated to the task of finding these suboptimal policies. The field of Approximate Dynamic Programming (ADP) considers these approaches (Powell, 2007), (Si, Barto, Powell, & Wunsch, 2004).

ADP, united with the field of reinforcement learning in (Bertsekas & Tsitsiklis, 1996), concerns itself with solving the Hamilton-Jacobi-Bellman (HJB) equation of dynamic programming. In discrete time, backwards induction is often used. For continuous time domains, the HJB equation takes on the form of a second-order partial differential equation. This work extends the HJB equation to the time scales calculus, therefore considering both discrete and continuous domains as well as mixed domains, which are neither discrete nor continuous, and formulates it using the very general alpha derivatives.

5.3 Dynamic Programming Overview

The requirements of the dynamic programming framework considered herein are as follows: a time scale $\mathbb{T}$ in which our decision points lie, controls $c(x(t),t)$, a stochastic disturbance $w(t)$, states $x(t)$ which evolve according to a rule $f(x(t),c(x(t),t),w(t),t)$, and a cost/reward $r(x(t),c,w(t),t)$ where the cost at a terminal decision point T is piecewise defined as $r_T(x(T))$. A policy π is a set of state-control pairs for each point in $\mathbb{T}$ such that each control is valid for both the state and time. Denote by π^t the tail of the policy π beginning with time step t. Also introduced is a cost-to-go function J given by

$$J_\pi(x(t_0),t_0) = E\left\{\int_{t_0}^{T} r(x(\tau),c(x(\tau),\tau),w(\tau),\tau)\Delta\tau\right\} \tag{5.1}$$

which measures the expected cost of a policy π. Assume that these expected values are finite and well defined.

Consider the following state-space dynamical system defined on a time scale $\mathbb{T}$:

$$x^{\alpha}(t) = f(x(t), c(x(t), t), w(t), t) \tag{5.2}$$

where $t \in \mathbb{T}$ and $x^{\alpha}(t)$indicates an interval taken using the alpha derivative. The task is to calculate a policy π which minimizes the cost-to-go function J_{π}. Call such a π an *optimal policy* and denote the optimal cost-to-go as $J^{*}(x(t_0), t_0) = \min_{\pi} J_{\pi}(\ x(t_0), t_0)$, where the minimum is considered over all policies.

Employ Bellman's Principle of Optimality in the solution to the optimization problem. This principle can be framed in the following way. Let π^{*} be an optimal policy. Then the optimal policy for the tail problem starting at time n, which is to minimize

$$E\left\{\int_{n}^{T} r(x(\tau), c(x(\tau), \tau), w(\tau), \tau)\Delta\tau\right\}, \tag{5.3}$$

is equal to the portion of π^{*}that overlaps π^{n}. To justify this principle, note that if it were not true, then the tail of π^{*}could be replaced by a more optimal π^{n}, thus contradicting the claim of optimality of π^{*}.

The dynamic programming algorithm, a form of backwards induction, involves stochastic optimization of control selection starting from the terminal time point T. Beginning with setting $J(x(T), T) = r_T(x(T))$, the algorithm proceeds via the following update rule:

$$\begin{aligned} J(x(t), t) &= \min_{c} E\{r(x(t), c(x(t), t), w(t), t) \\ &+ J(f(x(t), c((x(t), t), w(t), t), \sigma(t))\} \end{aligned} \tag{5.4}$$

for $t \in \mathbb{T}$. This rule says that the cost-to-go of the current state $x(t)$ under a control $c(x(t), t)$ equals the expected value of the immediate cost $r(x(t), c(x(t), t), w(t), t)$ plus the future costs $J(f(x(t), c((x(t), t), w(t), t), \sigma(t))$. The symbol $\sigma(t)$ represents the "next" point in our time scale $\mathbb{T}$, and the use of this forward-jump operator is one of the hallmarks of the time scales calculus's ability to combine discrete and continuous analysis.

It is the goal in this work to move beyond this dynamic programming algorithm and to establish the HJB equation in full on general domains using the alpha derivative equation.

5.4 Dynamic Programming Algorithm on Time Scales

Bellman's Principle of Optimality (Bertsekas, 2001), (Bertsekas & Tsitsiklis, 1996) aids in the solution of the above optimization problem. This principle can be stated in the following way. Let π^*be an optimal policy. Then the optimal policy for the tail problem starting at time n, which is to minimize

$$E\left\{\int_{n}^{T} r(x(\tau), c(x(\tau), \tau), w(\tau), \tau)\Delta\tau\right\}, \tag{5.5}$$

is equal to the portion of π^*that overlaps π^n. To justify this principle, note that if it were not true, then the tail of π^*could be replaced by a more optimal π^n, thus contradicting the claim of optimality of π^*.

The dynamic programming algorithm, a form of backwards induction, involves stochastic optimization of control selection starting from the terminal time point T. Beginning with setting $J(x(T), T) = r_T(x(T))$, the algorithm proceeds via the following update rule:

$$\begin{aligned} J(x(t), t) = \min_{c} E\{&r(x(t), c(x(t), t), w(t), t) \\ &+ J(f(x(t), c((x(t), t), w(t), t), \sigma(t))\} \end{aligned} \tag{5.6}$$

for $t \in \mathbb{T}$. This rule says that the cost-to-go of the current state $x(t)$ under a control $c(x(t), t)$ equals the expected value of the immediate cost $r(x(t), c(x(t), t), w(t), t)$ plus the future costs $J(f(x(t), c((x(t), t), w(t), t), \sigma(t))$. Recall that $\sigma(t)$ is the "next" point in our time scale $\mathbb{T}$.

It is standard to discuss optimality of this algorithm in terms that assume convergence. The proof contained herein, following Bertsekas, 2001, declares controls optimal if they minimize the update rule.

The classical version of this update rule is true for the discrete time scale $t = \{1, 2, \ldots, T\}$. This work extends this result to any isolated time scale $\mathbb{T}$.

5.4.1 Delta Derivative Version

This section provides the proof that the dynamic programming algorithm holds true for the case of the delta derivative on time scales.

Theorem 5.1 (Dynamic Programming Algorithm, Delta Derivatives). If $c^*(x(t), t)$ minimizes the update expression given above for each state and for all $t \in \mathbb{T}$, then the policy $c^*(x(t), t)$ is optimal.

Proof
Set $J^*(x(T), T) = r_T(x(T))$ and proceed via time scales induction to show that application of the dynamic programming algorithm's recursive update equations yields the optimal policy at each stage, i.e. that $J^*(x(t), t) = J(x(t), t))$ for all $t \in \mathbb{T}$.

Letting $t = T$ yields, by definition,

$$J^*(x(T),T) = r_T(x(T)) = J(x(T),T)) \tag{5.7}$$

Now, assume that $J^*(x(t),t) = J(x(t),t)$ for some time point $t \in \mathbb{T}$ and all states $x(t)$. To apply the backwards induction algorithm, recall that in a time scale, $\rho(t)$ is the point that comes just "before" the point t. Therefore, the quantity $\rho(t)$ plays a central role in our discussion, and the following equation gives the immediate one-step application of our update rule:

$$J^*(x(\rho(t)),\rho(t)) = \min_{c,\pi} E\left\{r(x(t),c(x(t),t),w(t),\rho(t)) + \int_t^T r(x(\tau),c(x(\tau),\tau),w(\tau),\tau)\Delta\tau\right\}. \tag{5.8}$$

The integral represents the value of the cost-to-go function J at the "next" time step after $\rho(t)$, which is t. Also note that the minimization is taken term-by-term over all controls and policies, respectively.

Use the principle of optimality to distribute the min through the expectation, as the tail problem is indeed an optimal policy for the problem contained within the tail. This yields the following:

$$\begin{aligned} J^*(x(\rho(t)),\rho(t)) = \min_{c} E &\left\{r(x(t),c(x(t),t),w(t),\rho(t))\right) \\ &+ \min_{\pi} E\left\{\left\{\int_t^T r(x(\tau),c(x(\tau),\tau),w(\tau),\tau)\Delta\tau\right\}\right\} \end{aligned} \tag{5.9}$$

Using the definition of $J^*(x(t),t)$, which subsumes the term minimized over the policy, we can reduce this expression to

$$J^*(x(\rho(t)),\rho(t)) = \min_{c} E\{r(x(t),c(x(t),t),w(t),\rho(t)) + J^*(x(t),t)\} \tag{5.10}$$

By the induction hypothesis, we know the optimal cost-to-go $J^*(x(t),t)$ is equivalent to the approximation $J(x(t),t)$ due to the dynamic programming algorithm. Thus, we write $J^*(x(\rho(t)),\rho(t))$ as

$$\min_{c} E\{r(x(t),c(x(t),t),w(t),\rho(t)) + J(x(t),t)\} \tag{5.11}$$

which, by definition, is simply

$$J^*(x(\rho(t)),\rho(t)) = J(x(\rho(t)),\rho(t)). \tag{5.12}$$

We have now satisfied conditions 1 and 2 of the principle of backwards induction on time scales given preliminarily. Since we assume $\mathbb{T}$ to be isolated, conditions 3 and 4 do not apply, and we conclude that, by backwards induction on time scales, we have proven our claim. ■

Thus, the dynamic programming algorithm is expanded to time scales. The computational requirements for implementing this algorithm, particularly for industrial-scale optimization problems common in operations research, are great (Powell, 2007). It is the task of ADP to calculate suboptimal policies in an efficient manner while simultaneously satisfying the needs of a given application. Within a time scales framework this approach is also valid, as the optimal update rule underlying the approximations holds.

5.4.2 *Quantum Calculus Version*

Set out with the following definitions: decision points $\boldsymbol{t}$ contained in a $\boldsymbol{q}$-time scale $\mathbb{T}$ with a terminal point $\boldsymbol{T}$, a set of controls for each state given by $\boldsymbol{c(x(t),t)}$, random disturbances modeled by a stochastic term $\boldsymbol{w(t)}$, a cost/reward function denoted by $\boldsymbol{r(x(t),c(x(t),t),w(t),t)}$ with terminal point T defined piecewise as $\boldsymbol{r_T(x(T))}$, and a dynamic system where states $\boldsymbol{x(t)}$ evolve according to the following rule:

$$d_q x(t) = f(x(t), c(x(t),t), w(t), t), \tag{5.13}$$

For policies π, require each set of state-control pairs to represent a valid association for both the space and time dimensions. The *tail* of the policy π, denoted as π^t, chronicles the sets of state-action pairs starting at decision point t and ending at the terminal point T. This notion is critical for discussion of the optimality principle and the subsequent derivation of the dynamic programming algorithm. The cost-to-go function is given by

$$\begin{aligned} J_\pi(x(t_0),t_0) &= E\,\langle \int_{t_0}^{q^T} r(x(\tau), c(x(\tau),\tau), w(\tau), \tau)\Delta\tau \rangle \\ &= (q-1)E\,\langle \sum_{i=t_0}^{q^{T-1}} q^i r(x(q^i), c(x(q^i),q^i), w(q^i), q^i) \rangle \end{aligned} \tag{5.14}$$

Define an *optimal policy* π^* to be one that minimizes the cost-to-go function J. The corresponding optimal cost-to-go function is denoted

$$J^*(x(t_0),t_0) = \min_\pi J_\pi(\,x(t_0),t_0) \tag{5.15}$$

where the *min* is considered over all policies. The goal of the dynamic programming problem is to calculate an optimal policy π^*. The most basic process by which this is achieved is called the *Dynamic Programming Algorithm*, which is a form of backwards induction. Starting from the terminal decision point T and following a schedule of recursively defined steps backwards in time towards the initial point t_0, the optimal policy can be calculated even in a stochastically rich environment. The algorithm begins with setting $J(x(T),T) = r_T(x(T))$ and proceeds via the following rule:

$$\begin{aligned} J(x(t),t) = \min_{c} E\langle r(x(t), c(x(t),t), w(t), t) \\ + J(f(x(t), c(t), w(t), t), q^{t+1})\rangle \end{aligned} \tag{5.16}$$

This recursion is a consequence of Bellman's Principle of Optimality, which states that any optimal policy must remain optimal when enacted on any tail of the system. That is, the solution that minimizes the cost-to-go function starting at any given point t,

$$(q-1)E\left\langle \sum_{i=t}^{q^{T-1}} q^i r(x(q^i), c(x(q^i), q^i), q^i)\right\rangle, \tag{5.17}$$

is simply the portion of the optimal policy π^* that coincides with the particular tail in question. The justification of this principle runs as a proof by contradiction: If the tail problem had a different solution than that given, then the cost-to-go function could be minimized further by changing out the optimal policy's tail with this alternate policy, thus prohibiting the optimal policy from being, in fact, optimal. This cannot be the case, so the optimality principle must hold.

The following proof follows that of Bertsekas, 2001 and suffices to establish the viability of dynamic programming in quantum calculus.

Theorem 5.2 (Dynamic Programming Algorithm in Quantum Calculus). The policy that minimizes the dynamic programming recursion (5.11) for all states and all times is optimal.

Proof: Set $J^*(x(T),T) = r_T(x(T))$ and proceed, via quantum calculus induction, to show that following the dynamic programming algorithm's update rule yields the optimal policy each step of the way, i.e. that $J^*(x(t),t) = J(x(t),t))$ for all $t \in \mathbb{T}$. Since the nature of this algorithm is to proceed backwards in time, the dual version of time scales induction as described in Section 4 will be used. (Recall that the backwards jump operator $\rho(t) = q^{t-1}$ for the q-time scale. However, we will maintain the use of the symbol $\rho(t)$ and trust no confusion will arise. This notation has the advantage of more closely mirroring the form of the version of the quantum calculus induction algorithm we invoke in the proof.)

Letting $t = T$ yields, by definition,

$$J^*(x(N), N) = r_T(x(T)) = J(x(N), N)). \tag{5.18}$$

Now, assume $J^*(x(t), t) = J(x(t), t)$ for some time point $t \in \mathbb{T}$ and all states $x(t)$. Then we have

$$\begin{aligned} J^*(x(\rho(t)), \rho(t)) &= r(x(t), c(x(t), t), w(t), \rho(t)) \\ &+ \min_{c,\pi} E\,\langle(q-1)E\,\langle \sum_{i=t_0}^{q^{T-1}} q^i r(x(q^i), c(x(q^i), q^i), w(q^i), q^i)\rangle\rangle\,. \end{aligned} \tag{5.19}$$

Note that the minimization is taken term by term over all controls and policies, respectively. We now use the principle of optimality to distribute the *min* through the expectation, as the tail problem is an optimal policy for the sub-problem contained within the tail. This yields the following:

$$\begin{aligned} J^*\left(x(\rho(t)), \rho(t)\right) &= \min_{c} E\langle r(x(t), c(x(t), t), w(t), \rho(t))\rangle \\ &+ \min_{\pi} E\,\langle(q-1)E\,\langle \sum_{i=t_0}^{q^{T-1}} q^i r(x(q^i), c(x(q^i), q^i), w(q^i), q^i)\rangle\rangle \end{aligned} \tag{5.20}$$

Using the definition of $J^*(x(t), t)$, which subsumes the term minimized over the policy, we can reduce this expression to

$$\begin{aligned} J^*(x(\rho(t)), \rho(t)) = \min_{c} E\langle r(x(t), c(x(t), t), w(t), \rho(t)) \\ + J^*(x(t), t)\rangle \end{aligned} \tag{5.21}$$

By the induction hypothesis, we know the optimal cost-to-go is equivalent to the approximation due to the dynamic programming algorithm. Thus, we write

$$\begin{aligned} J^*(x(\rho(t)), \rho(t)) = \min_{c} E\langle r(x(t), c(x(t), t), w(t), \rho(t)) \\ + J(x(t), t)\rangle \end{aligned} \tag{5.22}$$

which, by definition, is simply

$$J^*(x(\rho(t)), \rho(t)) = J(x(\rho(t)), \rho(t)) \tag{5.23}$$

which, in turn, is our desired result. ∎

With this, the dynamic programming algorithm is shown to work in quantum calculus. In fact, this can be interpreted as the Hamilton-Jacobi-Bellman equation in the quantum calculus, as the q-time scale is isolated. It should be noted, however, that this algorithm is quite computationally expensive, particularly for industrial-scale problems. To circumvent this failing, suboptimal methods are employed routinely. Collectively called Approximate Dynamic Programming (ADP), these algorithms seek to calculate suboptimal policies to whatever degree of accuracy is required by a given application. These techniques are tied quite intimately to backpropagation, and Chapter 6 of this book will provide a proof of the foundations of backpropagation in quantum calculus as well as in other time scales. In this way, both ADP and optimal dynamic programming are shown to have solid footing on q-time scales.

5.5 HJB Equation on Time Scales

Consider the dynamical system given by

$$x^{\Delta}(t) = f(x(t), c(t)) \tag{5.24}$$

where x represents states and c is the control. Let $t \in \mathbb{T}$, $x: \mathbb{T} \to \mathbb{R}$, and $x(\mathbb{T}) = \mathbb{T}_x$. The cost-to-go function $J: \mathbb{T}_x \times \mathbb{T} \to \mathbb{R}$ is given by

$$J(x(t_0), t_0) = \int_{t_0}^{T} r(x(t), u(t))\Delta t \tag{5.25}$$

where t_0 is the initial decision point and $r(x(t), u(t))$ is the cost. Assume J is delta-differentiable and x is σ_x-completely delta differentiable. Furthermore, require x to satisfy

$$x\big(\sigma(t)\big) = \sigma_x\big(x(t)\big). \tag{5.26}$$

Then the HJB equation on time scales is given by

$$\begin{aligned} 0 = \min_{c}\big(r\big(x(t), c(x(t), t)\big) &+ J^{\Delta_t}(x(t), t) \\ &+ J^{\Delta_x}(x(t), \sigma(t)) f(x(t), t)\big). \end{aligned} \tag{5.27}$$

This is an equation that any optimal policy of the minimization problem must satisfy. Since precious few industrial-scale applications admit an analytic solution of this equation, ADP is employed to develop approximation techniques for this purpose. The proof of this equation is the next theorem. The Hamilton-Jacobi equation is a result of the calculus of variations, and work extending this calculus to time scales is only just beginning (Atici, Biles, & Lebedinsky, 2006), (Bohner,

2004), (Bohner & Guseinov, 2007), (DaCunha, 2007), (DaCunha, 20085), (Ferreira & Torres, 2007), (Hilscher & Zeidan, 2004). These problems typically take the general form of minimizing the cost functional given by the following integral:

$$J(y) = \int_a^b L(t, y^\sigma(t), y^\Delta(t))\Delta t. \tag{5.28}$$

From this, the usual Euler and Legendre conditions can be derived on time scales. Our next result takes this a step further and proves the Hamilton-Jacobi equation for an alternate version, given by (5.25), of the above integral (5.28). Since equation (5.25) is the common cost functional of dynamic programming, the resulting equation is given the name Hamilton-Jacobi-Bellman. In this way, the following theorem is a contribution to the development of the calculus of variations on time scales as well as to ADP. This chapter proves the HJB equation for a form other than that given by (5.28), but there is still work to be done on Hamilton-Jacobi equations for more generalized cost functionals.

5.5.1 Delta Derivative Version

This section contains the proof of the Hamilton-Jacobi-Bellman equation on time scales using the usual delta derivative.

Theorem 5.3 (Hamilton-Jacobi-Bellman Equation). Let $V(x(t), t)$ be a solution to equation (5.25) such that

$$\begin{aligned} 0 = \min_c\big(r\big(x(t), c(x(t), t)\big) &+ V^{\Delta_t}(x(t), t) \\ &+ V^{\Delta_x}(x(t), \sigma(t)) f(x(t), t)\big). \end{aligned} \tag{5.29}$$

Assume the boundary conditions $V(x(T), T) = r_T(x(T))$ and $\hat{x}(t_0) = x(t_0)$, and suppose $c^*(x(t), t)$ attains the minimum called for in equation (17) for all states and all times. Let $x^*(t)$ be the state trajectory, subject to the condition $x^*(t_0) = x(t_0)$, that corresponds to applying the controls $c^*(x(t), t)$ at each decision point t.

Then the function $V(x(t), t)$ is the optimal cost-to-go function $J^*(x(t), t)$, and the control $c^*(x(t), t)$ is optimal.

Proof

Let $\hat{c}(x(t), t)$ be a control policy with corresponding state trajectory $\hat{x}(t)$. We will show that the policy $c^*(x(t), t)$ achieves a cost no greater than this arbitrary $\hat{c}(x(t), t)$, thus forcing $c^*(x(t), t)$ to be our optimal control. We begin by invoking equation (5.27) to give us

$$0 \le r\big(\hat{x}(t), \hat{c}(x(t), t)\big) + V^{\Delta_t}(\hat{x}(t), t) + V^{\Delta_x}(\hat{x}(t), \sigma(t)) f(\hat{x}(t), t). \tag{5.30}$$

Noting that, via (12), we have $x^{\Delta}(t) = f(x(t), c(t))$, we can rewrite (5.28) as

$$0 \leq r\big(\hat{x}(t), \hat{c}(x(t), t)\big) + V^{\Delta_t}(\hat{x}(t), t) + V^{\Delta_x}(\hat{x}(t), \sigma(t))x^{\Delta}(t). \tag{5.31}$$

By reversing the chain rule implicit in this formulation, we arrive at

$$0 \leq r\big(\hat{x}(t), \hat{c}(x(t), t)\big) + V^{\Delta}(t). \tag{5.32}$$

Integrating over our time horizon yields

$$0 \leq \int_{t_0}^{T} r(\hat{x}(t), \hat{c}(x(t), t))\Delta t + \int_{t_0}^{T} V^{\Delta}(t)\Delta t. \tag{5.33}$$

Using the fundamental theorem, we arrive at

$$0 \leq \int_{t_0}^{T} r(\hat{x}(t), \hat{c}(x(t), t))\Delta t + V(\hat{x}(T), T) - V(\hat{x}(t_0), t_0). \tag{5.34}$$

Substituting in our boundary conditions $V(x(T), T) = r_T(x(T))$ and $\hat{x}(t_0) = x(t_0)$ gives us

$$0 \leq \int_{t_0}^{T} r(\hat{x}(t), \hat{c}(x(t), t))\Delta t + r_T(x(T)) - V(x(t_0), t_0) \tag{5.35}$$

which is equal to

$$V(x(t_0), t_0) \leq \int_{t_0}^{T} r(\hat{x}(t), \hat{c}(x(t), t))\Delta t + r_T\big(x(T)\big). \tag{5.36}$$

From our hypothesis, we assume the controls $c^*(x(t), t)$ and their corresponding state trajectory $x^*(t)$ minimize the value function $V(x(t), t)$. Using this information and the initial condition $x^*(t_0) = x(t_0)$, we can replace the inequality with equality in the case of these quantities:

$$V(x(t_0), t_0) = \int_{t_0}^{T} r\big(x^*(t), c^*(x(t), t)\big)\Delta t + r_T\big(x^*(T)\big). \tag{5.37}$$

Combining with the previous equation, we have

$$\int_{t_0}^{T} r\big(x^*(t), c^*(x(t), t)\big)\Delta t + r_T\big(x^*(T)\big) \leq \int_{t_0}^{T} r\big(\hat{x}(t), \hat{c}(x(t), t)\big)\Delta t + r_T\big(x(T)\big). \tag{5.38}$$

This equation tells us that the cost of the policy $c^*(x(t),t)$ is less than or equal to the cost of any admissible policy $\hat{c}(x(t),t)$. We conclude the policy $c^*(x(t),t)$ is optimal and that, since $\hat{c}(x(t),t)$ is arbitrary, we have $V(x(t),t) = J^*(x(t),t)$. Therefore, any optimal policy must satisfy the HJB equation given by (15). ■

The calculus of time scales admits many different chain rules depending on various conditions on the functions of interest. The key step in the proof of the HJB equation is in the reversal of the chain rule. In principle, given any chain rule, a different form of the HJB equation can be derived. For example, the proof assumed the σ-complete differentiability of $x(t)$. If instead it is assumed that $x(t)$ is σ_x-completely differentiable, it is possible to obtain, by a different chain rule of Bohner and Guseinov, 2004, the following form of the HJB equation:

$$0 = \min_c \big(r\big(x(t), c(x(t),t)\big) + J^{\Delta_t}(\sigma_x(x(t)),t) + J^{\Delta_x}(x(t),t) f(x(t),t)\big) \tag{5.39}$$

This difference could prove crucial in an application, depending on the form of the state variable $x(t)$. It is important to note that further research on time scales calculus into new and more powerful versions of the chain rule for partial derivatives will result in new ways to frame the Hamilton-Jacobi-Bellman equation on time scales.

5.5.2 Nabla Derivative Version

Now we extend the theorem proven in the previous section for the case of the backwards, or nabla, time scales derivative. Many of the details remain similar. This content appeared in Seiffertt, Sanyal, & Wunsch, 2008b.

Consider now the system

$$x^\nabla(t) = f(x(t), u(t)) \tag{5.40}$$

where x represents states and u is the control. Let $t \in \mathbb{T}$, $x: \mathbb{T} \to \mathbb{R}$, and $x(\mathbb{T}) = \mathbb{T}_x$. The cost-to-go function $J: \mathbb{T}_x \times \mathbb{T} \to \mathbb{R}$ is given by

$$J(x(t_0), t_0) = \int_{t_0}^{T} r(x(t), u(t)) \nabla t \tag{5.41}$$

where t_0 is the initial decision point and $r(x(t), u(t))$ is the cost. Assume J is delta-differentiable and x is ρ_x-completely delta differentiable. Furthermore, require x to satisfy

$$x\big(\rho(t)\big) = \rho_x\big(x(t)\big). \tag{5.42}$$

Then the HJB equation on time scales is given by

$$0 = \min_{c}\Big(r\big(x(t), u(x(t), t)\big) + J^{\nabla_t}(x(t), t) + J^{\nabla_x}(x(t), \rho(t))f(x(t), t)\Big). \tag{5.43}$$

This is an equation that any optimal policy of the minimization problem must satisfy.

Theorem 5.4 (Hamilton-Jacobi-Bellman Equation with Backwards Derivatives). Let $V(x(t), t)$ be a solution to equation (15) such that

$$0 = \min_{c}\Big(r\big(x(t), u(x(t), t)\big) + V^{\nabla_t}(x(t), t) + V^{\nabla_x}(x(t), \rho(t))f(x(t), t)\Big). \tag{5.44}$$

Assume the boundary condition $V(x(T), T) = r_T(x(T))$ and $\hat{x}(t_0) = x(t_0)$ and suppose $u^*(x(t), t)$ attains the minimum called for in equation (5.54) for all states and all times. Let $x^*(t)$ be the state trajectory, subject to the condition $x^*(t_0) = x(t_0)$, that corresponds to applying the controls $u^*(x(t), t)$ at each decision point t.

Then the function $V(x(t), t)$ is the optimal cost-to-go function $J^*(x(t), t)$, and the control $u^*(x(t), t)$ is optimal.

Proof
Let $\hat{u}(x(t), t)$ be a control policy with state trajectory $\hat{x}(t)$. Our goal is to show that the policy $u^*(x(t), t)$ achieves a cost equal to, at most, this arbitrary $\hat{u}(x(t), t)$, which will mean that $u^*(x(t), t)$ is our optimal control. We begin using equation (5.44) to give

$$0 \le r\big(\hat{x}(t), \hat{u}(x(t), t)\big) + V^{\nabla_t}(\hat{x}(t), t) + V^{\nabla_x}(\hat{x}(t), \rho(t))f(\hat{x}(t), t). \tag{5.45}$$

Noting that, via (10), we have $x^{\nabla}(t) = f(x(t), u(t))$, we can rewrite (5.45) as

$$0 \le r\big(\hat{x}(t), \hat{u}(x(t), t)\big) + V^{\nabla_t}(\hat{x}(t), t) + V^{\nabla_x}(\hat{x}(t), \rho(t))x^{\nabla}(t). \tag{5.46}$$

and, by using the chain rule, we can rewrite as

$$0 \le r\big(\hat{x}(t), \hat{u}(x(t), t)\big) + V^{\nabla}(t). \tag{5.47}$$

Integrating over the time horizon yields

$$0 \leq \int_{t_0}^{T} r(\hat{x}(t), \hat{u}(x(t), t))\nabla t + \int_{t_0}^{T} V^{\nabla}(t)\nabla t. \tag{5.48}$$

Using the fundamental theorem, substituting boundary conditions, and rearranging terms gives

$$V(x(t_0), t_0) \leq \int_{t_0}^{T} r(\hat{x}(t), \hat{u}(x(t), t))\nabla t + r_T(x(T)). \tag{5.49}$$

The details of these operations follow those in the proof of Theorem 5.2. From our hypothesis, we assume the controls $u^*(x(t), t)$ and their corresponding state trajectory $x^*(t)$ minimize the value function $V(x(t), t)$. Using this information and the initial condition $x^*(t_0) = x(t_0)$, we can replace the inequality with equality in the case of these quantities:

$$V(x(t_0), t_0) = \int_{t_0}^{T} r(x^*(t), u^*(x(t), t))\nabla t + r_T(x^*(T)). \tag{5.50}$$

Combining with the previous equation, we have

$$\begin{aligned} &\int_{t_0}^{T} r(x^*(t), u^*(x(t), t))\nabla t + r_T(x^*(T)) \\ &\qquad \leq \int_{t_0}^{T} r(\hat{x}(t), \hat{u}(x(t), t))\nabla t + r_T(x(T)). \end{aligned} \tag{5.51}$$

This equation says that the cost of the policy $u^*(x(t), t)$ is less than or equal to the cost of any admissible policy $\hat{u}(x(t), t)$. We conclude the policy $u^*(x(t), t)$ is optimal and that, since $\hat{u}(x(t), t)$ is arbitrary, we have $V(x(t), t) = J^*(x(t), t)$. Therefore, any optimal policy must satisfy the HJB equation given by (14). ■

5.5.3 Alpha Derivative Version

The final derivation of the HJB equation is proven to be viable on the most general of the time scales derivatives, thus showing that the fundamentals of dynamic programming hold for domains vastly more general than any considered before this point.

Consider now the dynamical system given by

$$x^{\alpha}(t) = f(x(t), c(t)) \tag{5.52}$$

where x represents the states and c represents the control. Let $t \in \mathbb{T}$, $x: \mathbb{T} \to \mathbb{R}$, and $x(\mathbb{T}) = \mathbb{T}_x$. The cost-to-go function $J: \mathbb{T}_x \times \mathbb{T} \to \mathbb{R}$ is given in the alpha case by

$$J(x(t_0), t_0) = \int_{t_0}^{T} r(x(t), c(t))\alpha t \tag{5.53}$$

where t_0 is the initial decision point and $r(x(t), c(t))$ is the cost.

Then the HJB equation on generalized time scales is given by

$$\begin{aligned} 0 = \min_c \big(r\big(x(t), c(x(t), t)\big) &+ J^{\alpha_t}(x(t), t) \\ &+ J^{\alpha_x}(x(t), \alpha(t)) f(x(t), t)\big). \end{aligned} \tag{5.54}$$

This is an equation that any optimal policy of our minimization problem must satisfy. The following theorem states the HJB equation with the alpha derivative.

Theorem 5.5 (Hamilton-Jacobi-Bellman Equation Using Alpha Derivatives). Let $V(x(t), t)$ be a solution to equation (5.53) such that

$$\begin{aligned} 0 = \min_c \big(r\big(x(t), c(x(t), t)\big) &+ V^{\alpha_t}(x(t), t) \\ &+ V^{\alpha_x}(x(t), \alpha(t)) f(x(t), t)\big). \end{aligned} \tag{5.55}$$

Assume the boundary condition $V(x(T), T) = r_T(x(T))$ and $\hat{x}(t_0) = x(t_0)$, and suppose $c^*(x(t), t)$ attains the minimum called for in equation (14) for all states and all times. Let $x^*(t)$ be the state trajectory, subject to the condition $x^*(t_0) = x(t_0)$, that corresponds to applying the controls $c^*(x(t), t)$ at each decision point t.

Then the function $V(x(t), t)$ is the optimal cost-to-go function $J^*(x(t), t)$, and the control $c^*(x(t), t)$ is optimal.

The proof of this theorem is similar to that of Theorems 5.2 and 5.3. It appears in full in Seiffertt, 2009.

5.6 Conclusions

The time scales calculus is an increasingly relevant and developed area of mathematics with wide-ranging opportunities for application. This book has established that the dynamic programming algorithm, derived from Bellman's Principle of Optimality, pertains to time scales. Also derived is the Hamilton-Jacobi-Bellman equation on time scales, and it has been demonstrated that a family of such equations exist. The solution of such an equation is the fundamental goal of ADP. This chapter identifies three significant directions that the investigation of ADP on time scales can take. First, as the derivation of the HJB equation was dependent on the mathematics of the time scales calculus of multiple variables in general, and on the chain rule in particular, further variations and extensions in this area will prove critical. The generalized Stokes theorem, principles of the variational calculus, and more complete chain rules are three areas where new contributions are of exceptional need.

Second, numerical approximation work in time scales remains a promising endeavor. With the availability of computational resources such as the Time Scales MatLab Toolbox from the Baylor University Time Scales Group, both applied and theoretical investigation into the numerics of time scales calculus can be pursued. Numerical differentiation and integration techniques on time scales would provide significant value, as would time scales extensions of optimization algorithms, be they population-based models from the computational intelligence literature or provably convergent methods from applied mathematics (Abramson & Audet, 2006). Also needed are demonstrations of ADP-based controllers operating in a time scales framework. This brings us to our third direction for growth: applications.

In addition to the electric circuit, population biology, and virus outbreak modeling applications to which time scales has been applied (Bohner & Peterson, 2001), the field of time scales control needs to show significant upgrade to larger-scale problems. Analysis of technical trading rules, macroeconomic dynamical models, and monetary policy are areas in economics and finance in which time scales can be used and in which time scales-based controllers would be of great interest.

While the study of time scales can provide a concise theoretical unification of control theory in the discrete and continuous case, it can also provide so much more than that. There are important application areas in which dealing simultaneously with discrete and continuous variables is critical (Werbos, 2006), and the time scales calculus provides a natural and powerful framework for such exploration.

Chapter 6
Backpropagation on Time Scales

6.1 Overview

This section extends the previous section's focus on the unified computational intelligence goal of developing the capability to *adapt*. The dynamic programming algorithm typically utilizes neural networks as function approximation tools. Therefore, discussing how to train a neural network within the unified framework of the time scales calculus contributes directly to this goal.

The results presented here are adapted from a paper that has been accepted for publication in IEEE Transactions on Neural Networks but that, at the time of this book's completion, has not yet appeared.

6.2 Introduction

Backpropagation, based on the mathematical notion of an ordered derivative, is the most widely used neural network learning technique.. This section presents a formulation of ordered derivatives and the backpropagation training algorithm using the important emerging area of mathematics known as the time scales calculus. This calculus, with its potential for application to a wide variety of inter-disciplinary problems, is becoming a key area of mathematics. It is capable of unifying continuous and discrete analyses within one coherent, theoretical framework. Using this calculus, this book presents a generalization of backpropagation appropriate for cases beyond the specifically continuous or discrete. This book also develops a new multivariate chain rule of this calculus, defines ordered derivatives on time scales, proves a key theorem about them, and derives the backpropagation weight update equations for a feedforward, multi-layer neural network architecture. By drawing together the time scales calculus and the field of neural network learning, the first connection between these two areas is presented.

The desire to handle both discrete and continuous input signals within one theoretical framework has been driving much recent research in areas such as optimal control, the modeling of dynamic ecological population systems, intelligent robotic systems, and economic utility theory. Many applications require such a capability, and the emerging area of mathematics known as the time scales

J. Seiffertt & D.C. Wunsch: Unified Computational Intell. for Complex Sys., ALO 6, pp. 77–89.
springerlink.com

calculus is being used increasingly for such purposes. Introduced in (Hilger, 1989), the time scales calculus provides a unified language for handling both discrete and continuous equations.

A model of the backpropagation algorithm based on this time scales calculus is presented. The variables are allowed to take on either discrete or continuous values, and a rule to update the weights in this general form is derived. The original version of backpropagation was discrete in nature, but researchers have since developed it in the continuous case. These results are the first utilizing this new mathematics to discuss neural network learning in a more general way. No restriction is made in this development regarding whether the network variables ought to be discrete or continuous. Models admitting either difference or differential equations can be discussed fruitfully without our framework.

Ordered derivatives, a concept invented by Paul Werbos to track the sensitivities of variables within the sorts of systems often studied by social scientists, are also treated in this general time scales case. Key to the development of the backpropagation algorithm is the notion of a chain rule for ordered derivatives. This chain rule in the time scales calculus is proven, and what an ordered derivative means within this new mathematics is defined. Note that the time scales calculus itself admits a wide variety of derivatives. Therefore, care must be taken when working with chain rules and differentiation in general. These results hold for the common delta derivative and its backwards analog, the nabla derivative. Further derivatives, including the alpha and diamond-alpha derivatives, may also be applied to gradient descent-based neural network learning in the future.

This chapter is organized as follows. The reader is referred to Chapter 4 of this book, which covers basic notions of the time scales calculus needed for our work here. It contains the proof of a new multi-variable chain rule on time scales. Section 6.3 presents the ordered derivatives, discusses network conventions, and presents the proof of the ordered derivative chain rule. Section 6.4 uses this theorem to derive the backpropagation update equations on time scales. Finally, Section 6.5 discusses the limitations and extensions of the work and provides a view of the future of this avenue of research.

6.3 Ordered Derivatives

In his PhD dissertation (Werbos, 1974, reprinted in Werbos, 1994), Paul Werbos introduced the idea of an *ordered derivative* distinct from the notion of a traditional direct partial derivative in common use in the analysis of physical systems. The use of these ordered derivatives is motivated by the particular character possessed by social systems as compared to those systems that operate under strictly physical dynamics. In these systems, it makes sense to model using a series of variables that are each the consequence of a variable appearing earlier in the system. Hence, these are called *ordered variables*.

Werbos shows that the traditional chain rule for total derivatives fails to provide an adequate framework for the calculations needed for the investigation not only of such an integrated socio-economic system, but also for a wide array of other mathematically related connectionist systems, such as the problem of political forecasting described in detail in his dissertation work. However, he goes on to prove that a chain rule for ordered derivatives is in fact capable of driving the necessary calculation engine for studying these sorts of systems. While he does not term this method *backpropagation,* that being the name given the technique when popularized and independently discovered by others, the algorithm he lays out is in actuality the very same one which today powers most of the neural network learning worldwide. In addition, Werbos establishes backpropagation as a viable manner by which to calculate the needed derivatives not only in neural network training but for the manipulation of weights for a broad spectrum of adaptive systems.

To complete the theoretical foundation of backpropagation, Werbos proved a special chain rule for these ordered derivatives that worked for the applications in which the traditional chain rule of continuous analysis broke down. This chapter contains the definition of the ordered (delta) derivative on time scales and the proof that the chain rule obtained by Werbos holds true in this new environment.

6.3.1 Network Definitions

Following Werbos, define the following system variables: input $X(t)$ and output $\bar{Y}$, which approximates the target output $Y(t)$. The dimension of the outputs is denoted by n. The relationship between inputs and outputs is determined via a series of adaptive weights W. It is the goal of backpropagation, and hence the ordered derivative calculations, to tune W in such a way as to reduce an error measure between $\bar{Y}$ and Y. The implicit time scale in Werbos's work is an isolated subset of the integers $\mathbb{Z}$. This book considers time scales of higher generality and will assume $t \in \mathbb{T}$, where $\mathbb{T}$ is not restricted to a subset of $\mathbb{Z}$. Let N denote the number of inputs or the number of time steps the system is allowed to compute.

Define the following measure of system performance/error:

$$J = \int_1^{\sigma(N)} J(\tau)\Delta\tau = \int_1^{\sigma(N)} \sum_{i=1}^{n} \frac{1}{2}(\bar{Y}_i(\tau) - Y_i(\tau))^2 \Delta\tau \tag{6.1}$$

This form represents the standard least-squares error measure. Note that the integral takes on the role of the generalized summation operation and is appropriate for any time scale $\mathbb{T}$ free of restriction to an isolated case. The proper summation that remains represents the dimension of the output vector, which is a fixed scalar value not related to our time scale $\mathbb{T}$ and is thus not subsumed by an integral.

The form of this measure J when defined on different time scales can be investigated.

Consider a function $f:\mathbb{T}\to\mathbb{R}$ where $\mathbb{T}=\{t_0,t_1,\ldots\}$. Then, $\int_{t_0}^{t} f(\tau)\Delta\tau = \sum_{\tau\in[t_0,t)} f(\tau)\mu(\tau)$, where t_0 is the starting point of the time scale $\mathbb{T}$ and μ is the graininess function of $\mathbb{T}$ as defined in Chapter 4. So, if $\mathbb{T}=h\mathbb{Z}$, then the performance measure J becomes

$$\begin{aligned} J &= \int_{t_0}^{N} J(\tau)\Delta\tau\sigma = \sum_{t\in[t_0,t)} J(\tau)h \\ &= h\sum_{t\in[t_0,t)}\sum_{i=1}^{n}\frac{1}{2}(\overline{Y}_i(\tau)-Y_i(\tau))^2 \end{aligned} \tag{6.2}$$

Similarly, for the case of the quantum calculus time scale $q^{\mathbb{N}_0}$, $q>1$, the following error obtains:

$$\begin{aligned} J &= \int_{t_0}^{q^n} J(\tau)\Delta\tau = \sum_{i=0}^{n-1} J(q^i)q^i(q-1) \\ &= (q-1)\sum_{i=0}^{n-1}\left(\frac{q^i}{2}\right)(\overline{Y}_i(q^i)-Y_i(q^i))^2 \end{aligned} \tag{6.3}$$

In principle, this error term can be calculated for any time scale under consideration. It should be mentioned that in the case in which the time scale $\mathbb{T}$ is not isolated, the resulting integral will be of the form $\int_{t_0}^{\sigma(t)} J(\tau)\Delta\tau$, where the upper limit is the forward jump operation applied to the variable rather than simply the variable itself. This requirement follows from technical details of integration theory on time scales beyond the scope of the current discussion. The theory presented thus far provides sufficient background for the theorems presented in this book.

Regardless of the choice of time scale, which informs the construction of the error measure, the network equations remain to be analyzed. The chain rule for ordered derivatives is needed here, as this is the tool used to turn the network equations into a computation unit for the proper updating of the adaptive weights via backpropagation. As a preamble to the chain rule theorem, it is first necessary to discuss the notion of ordered derivatives.

6.3.2 Structure of Ordered Derivatives

In order to discuss the structure of ordered derivatives, consider the following system:

$$\begin{aligned} x_4 &= 2x_1 + 5x_2 + x_3 \\ x_3 &= 4x_2 + x_1 \\ x_2 &= 7x_1 \end{aligned}$$

Distinguish between *direct* and *indirect* effects. The direct effect of a change in x_1 on x_4, denoted $\frac{\partial x_4}{\partial x_1}$, is equal to 2. Note that in this case the ordering of the variables is not taken into account. Unlike a physical system in which the evaluations of all the variables x_i would be assumed to take place simultaneously so that the calculation of $\frac{\partial x_4}{\partial x_1}$ would require the substitution of lower valued x_i's into the equation for x_4, in this *ordered system* the direct effect is calculated wholly from the equation for x_4 without recourse to the equations representing lower, or previous, levels of activity. The total derivative effect, then, in such an ordered system is called the indirect effect and is denoted by $\frac{\partial^+ x_4}{\partial x_1}$ In this case, this indirect effect is equal to 66.

Furthermore, this notation corresponds to the continuous case. For the general time scale case, denote the ordered derivative of the performance measure J with respect to an ordered variable x_i as

$$J^{\Delta_{x_i}^{+}}$$

The following text develops the necessary definitions to make the above notation precise.

Let $x_1(t), x_2(t), \ldots, x_n(t)$ be an ordered sequence of variables with $t \in \mathbb{T}$. These variables represent stages of a larger calculation (e.g., layers, in a sense of a multi-layer perceptron network). What sets them apart as *ordered* is that they follow a recursion given by

$$x_i = f(x_1, x_2, \ldots, x_{i-1}) \tag{6.4}$$

In this way, the mathematics speak meaningfully of causation as a basis for the relationship among the x_i's. The interest is in determining the way in which the performance term J changes with respect to one of the x_i's, i.e., it is desired to calculate $J^{\Delta x_i}$. Following Werbos, set up the error, which itself is actually the final variable in the extended ordered set, as a sequence of recursive functions, which is required to be σ_1-completely (delta) differentiable, such that

$$J_n(x_1, \ldots, x_n) = x_n \tag{6.5}$$

and

$$J_{i-1}(x_1, \ldots, x_{i-1}) = J_i(x_1, \ldots, x_{i-1}, f_i(x_i, \ldots, x_{i-1})) \tag{6.6}$$

These recursive functions describe the final node in the network, x_n, as a function of earlier variables in the ordering successively calculated. These functions J_i

will provide the machinery needed to discuss the chain rule for ordered derivatives. In application, J_n, also equal to the final variable in the ordering x_n, is the network performance/value measure that is to be optimized via the updating of adaptive weights.

Now, the performance can be defined by reference to these recursive J_i functions. Define the ordered (delta) derivative of x_n with respect to a previous variable x_i in the ordering by

$$x_n^{\overset{+}{\Delta} x_i} = J_i^{\Delta x_i} \tag{6.7}$$

where the functional relationship among the x_i's is repressed (i.e., all variables $x_1, x_2, \dots, x_{i-1}$ are held constant) when calculating $J_i^{\Delta x_i}$ so as to properly represent the ordered nature of these variables (that is, the derivative $J_i^{\Delta x_i}$ gives us only the *direct* effect of a change in x_i on J_i.)

With these definitions, the following chain rule may be proven.

6.3.3 The Chain Rule

The chain rule for ordered derivatives extends the normal chain rule of traditional calculus to the special case of variables within an ordered system. In this way, the full interplay among the components of such a system can be realized from the mathematics. The backpropagation algorithm of derivative calculation hinges on sifting the network equations given in the previous section through the ordered derivative. In particular, this chain rule is key.

Theorem 6.1 (The Chain Rule for Ordered Derivatives)

$$J_j^{\Delta x_i} = \sum_{k=j+1}^{n} x_n(\sigma_1(x_1), x_2, \dots, x_{n-1})^{\overset{+}{\Delta} x_k} x_k^{\Delta x_i} \tag{6.8}$$

Proof

We proceed via reverse induction on the index j. We will start with $j = n - 1$ and end with $j = 1$. Note that with $j < i$, the nature of the definitions of the f_i's and x_i's force the terms $f_k^{\Delta x_i}$ to equal 0; therefore, these terms do not contribute to the summation. So, it will suffice to consider j in the range $n - 1$ to i.

Let $j = n - 1$. Then our hypothesis becomes

$$J_n^{\Delta x_i} = x_n(\sigma_1(x_1), x_2, \dots, x_{n-1})^{\overset{+}{\Delta} x_k} x_n^{\Delta x_i} = x_n^{\Delta x_i} \tag{6.9}$$

Since by definition $J_n = x_n$, we have an identity and the claim is proven.

Now assume our hypothesis holds for $j+1<n$. We will proceed to show it is true for $j>i$. Noting that the recursive definition of the J_i's make them all equal, we have that $J_j^{\Delta x_i} = J_{j+1}^{\Delta x_i}$. This expression now requires use of our chain rule for n variables proven in Chapter 4. Write J_{j+1} explicitly as a function of x_i:

$$J_{j+1}\left(f_1(x), \ldots, f_{j-1}(x_1, \ldots, x_{j-1}), f_{j+1}(x_1, \ldots, x_j)\right) \tag{6.10}$$

Due to the nature of the ordered system, $x_k^{\Delta x_i} = 0$ for $i \neq 0$ and $i < j$ so that the expansion given by our n variable chain rule

$$J_{j+1}^{\Delta x_i} = J_{j+1}^{\Delta x_1} x_1^{\Delta x_i} + \sum_{k=2}^{j+1} J_{j+1}(\sigma_1(x_1), x_2, \ldots, x_n)^{\Delta x_k} x_k^{\Delta x_i} \tag{6.11}$$

reduces to

$$J_{j+1}^{\Delta x_i} = J_{j+1}^{\Delta x_{j+1}} x_{j+1}^{\Delta x_{j+1}} + J_{j+1}^{\Delta x_i} x_i^{\Delta x_i} \tag{6.12}$$

so that

$$J_j^{\Delta x_i} = J_{j+1}^{\Delta x_{j+1}} x_{j+1}^{\Delta x_i} + J_{j+1}^{\Delta x_i} \tag{6.13}$$

Our induction hypothesis gives us

$$J_{j+1}^{\Delta x_i} = \sum_{k=j+2}^{n} x_n(\sigma_1(x_1), x_2, \ldots, x_{n-1})^{\overset{+}{\Delta} x_k} x_k^{\Delta x_i} \tag{6.14}$$

Combining these results yields

$$J_{j+1}^{\Delta x_i} = J_{j+1}^{\Delta x_{j+1}} x_{j+1}^{\Delta x_i} + \sum_{k=j+2}^{n} x_n(\sigma_1(x_1), x_2, \ldots, x_{n-1})^{\overset{+}{\Delta} x_k} x_k^{\Delta x_i} \tag{6.15}$$

which is equal to

$$x_n^{\overset{+}{\Delta} x_{j+1}} x_{j+1}^{\Delta x_i} + \sum_{k=j+2}^{n} x_n(\sigma_1(x_1), x_2, \ldots, x_{n-1})^{\overset{+}{\Delta} x_k} x_k^{\Delta x_i} \tag{6.16}$$

and which reduces to

$$\sum_{k=j+2}^{n} x_n(\sigma_1(x_1), x_2, \dots, x_{n-1})^{\Delta_{x_k}^{+}} x_k^{\Delta x_i} \tag{6.17}$$

Thus, our claim, the chain rule for ordered derivatives, is proven. ■

The usual and most useful form of this theorem is given by the following corollary:

Theorem 6.2 (Chain Rule for Ordered Derivatives Corollary)

$$\begin{aligned} x_n(\sigma_1(x_1), x_2, \dots, x_{n-1})^{\Delta_{x_i}^{+}} &= x_n(\sigma_1(x_1), x_2, \dots, x_{n-1})^{\Delta x_i} \\ &+ \sum_{k=i+1}^{n-1} x_n(\sigma_1(x_1), x_2, \dots, x_{n-1})^{\Delta_{x_i}^{+}} x_k^{\Delta x_i} \end{aligned} \tag{6.18}$$

This corollary follows immediately from the chain rule theorem. Note that in practice, J is used as the final variable in the ordered system, x_n, so that this corollary will take the form

$$J^{\Delta_{x_i}^{+}} = J^{\Delta x_i} + \sum_{k=i+1}^{n-1} J^{\Delta_{x_k}^{+}} x_k^{\Delta x_i} \tag{6.19}$$

Furthermore, while this construction uses the delta derivative, the nabla derivative formulation likewise follows immediately from the delta:

$$J^{\nabla_{x_i}^{+}} = J^{\nabla x_i} + \sum_{k=i+1}^{n-1} J^{\nabla_{x_k}^{+}} x_k^{\nabla x_i} \tag{6.20}$$

The nabla derivative formulation is not used to derive backpropagation, but a full theory of neural network learning could indeed be built upon this derivative. This is a direction for future research.

Note that the time scales version of the chain rule for ordered derivatives contains the term $\sigma_1(x_1)$. This is due to the requirement of σ_1-complete (delta) differentiability in our n-variable chain rule. Further versions of both chain rules can be developed which require σ_i differentiability, where $i \neq 1$.

With this chain rule, the weight update rules for the backpropagation algorithm for neural network learning can be derived.

6.4 The Backpropagation Algorithm on Time Scales

Prior to the development of backpropagation as the method of choice for calculating derivatives to update the weights of neural networks, a direct differentiation method was used. This method demands much in terms of computational resources when the number of weights is larger than the number of neurons. Since multi-layer feedforward systems typically satisfy this criterion, the more efficient backpropagation is preferred. For a derivation of this algorithm in the traditional continuous case, the reader is directed to a standard neural network reference text such as Principe, Euliano, and Lefebvre, 2000.

To derive the backpropagation algorithm on time scales, work from the assumption that the node activity of our network takes on the following form:

$$x_i = f\left(\sum_{j<i} w_{ij} w_j\right) \tag{6.21}$$

where the w_{ij} are the adaptive weights. Note that these weights are part of the ordered system model for the neural network.

From this, the change in an ordered variable with respect to changes in an earlier node using the time scales chain rule can be calculated:

$$x_j^{\Delta x_i} = w_{ij} \int_0^1 f'\left(\sum_{j>i} w_{ij} x_j + h\mu(t) w_{ij}\right) dh \tag{6.22}$$

Now, the rate of change of the error measure J with respect to an activity node x_i is given by

$$J^{\Delta \overset{+}{x}_i} = -J^{\Delta x_i} + \sum_{k=i+1}^{n-1} J^{\Delta \overset{+}{x}_k} x_j^{\Delta x_i} \tag{6.23}$$

which via the chain rule may be written as

$$J^{\Delta \overset{+}{x}_i} = -J^{\Delta x_i} + \sum_{k=i+1}^{n-1} J^{\Delta \overset{+}{x}_k} w_{ij} \int_0^1 f'\left(\sum_{j>i} w_{ij} x_j + h\mu(t) w_{ij}\right) dh \tag{6.24}$$

and reduced to

$$J^{\Delta \overset{+}{x}_i} = -J^{\Delta x_i} + \sum_{k=i+1}^{n-1} w_{ij} \delta_i \tag{6.25}$$

where

$$\delta_i = J^{\Delta_{x_k}^+} \int_0^1 f'\left(\sum_{j>i} w_{ij}x_j + h\mu(t)w_{ij}\right) dh \tag{6.26}$$

is the *local error* term, which measures the change in the performance measure with respect to a change at the level of a local neuron.

The gradient descent *delta rule* on time scales is given by

$$w_{ij}^{\Delta}(t) = -\beta\delta_i x_j(t) \tag{6.27}$$

which captures the change in the weights w_{ij} as a change in the error function J modified by a learning rate β.

The final step in the derivation of the backpropagation algorithm is the integration of the delta learning rule with the equations obtained previously:

$$w_{ij}^{\Delta}(t) = -\beta J^{\Delta_{w_{ij}}^+} \tag{6.28}$$

This form, which holds for nodes such that $x\colon \mathbb{T} \to \mathbb{R}$, can account for networks in which the input is either continuous or discrete using a single theoretical framework and update equation. This also allows for the construction of connectionist systems capable of processing inputs that can switch between continuous and discrete signals while actively calculating. Further extensions of the time scales calculus to unifications in the neural networks field are discussed in the conclusion below.

6.5 Quantum Calculus

This section formulates the ordered derivative in quantum calculus and proves that the chain rule derived by Werbos pertains to this alternate environment. It concludes that the backpropagation approach to q-derivative calculation is as valid as the one for the classical derivative and, as such, neural network training on quantum calculus may follow its traditional counterpart.

Further analysis of the network equations requires the chain rule for ordered derivatives, as this is the tool used to transform the network equations into a calculator of the proper updates of a system's adaptive weights via backpropagation. As a preamble to the chain rule theorem, it is necessary to define what is meant by an ordered derivative for the quantum calculus environment.

Let $x_1(t), x_2(t) \dots, x_n(t)$ be an ordered sequence of variables with $t \in \mathbb{T}$. These variables represent stages of a larger calculation (e.g., layers, in a sense, of a multi-layer perceptron network) and follow a recursion given by

$$x_i = f_i(x_{i-1}, x_{i-2}, \dots, x_1) \tag{6.29}$$

so that causation as a basis for the relationships among the x_i's can be analyzed. As with the case considered previously, it is important to determine the way the error E changes with respect to one of the x_i's. Following Werbos, 1994, set up the error as a sequence of recursive functions such that

$$E_n(x_n, \dots, x_1) = x_n \tag{6.30}$$

and

$$E_{i-1}(x_{i-1}, \dots, x_1) = E_i(f_i(x_{i-1}, \dots, x_1), x_{i-1}, \dots, x_1). \tag{6.31}$$

Then, the ordered derivative of E, which equals x_n, is defined to be

$$E^{d_q^+ x_i} = \frac{d_q E_i}{d_q x} = E_i^{d_q x_i}. \tag{6.32}$$

The backpropagation algorithm of derivative calculation hinges on sifting the network equations through the ordered differentiation operator. In particular, the chain rule for ordered derivatives plays a key role. The following theorem establishes this chain rule for the quantum calculus.

Theorem 6.3 (Ordered Derivative Chain Rule in the Quantum Calculus)

$$E_j^{d_q x_i} = \sum_{k=j+1}^{n} x_n^{d_q^+ x_k} f_k^{d_q x_i}$$

Proof

As in Werbos, 1994, we proceed by induction on j. We will start with $j = n-1$ and end with $j = i$. With $j < i$, the recursive definitions of the f_i's and x_i's force the terms $f_k^{d_q x_i}$ to zero; therefore, they do not contribute to the summation. So, it will suffice to consider j in the range $n-1$ to i.

Let $j = n-1$. Then our hypothesis becomes

$$E_{n-1}^{d_q x_i} = x_n^{d_q^+ x_{n-1+1}} f_{n-1+1}^{d_q x_i} = x_n^{d_q^+ x_n} f_n^{d_q x_i}. \tag{6.33}$$

Calling on the definition of the sequence of E_i's, we see that $E_{n-1}^{d_q x_i} = f_n^{d_q x_i}$ so that, since $x_n^{d_q^+ x_n} = 1$, the claim is proven.

Now, assume the hypothesis is true for some $j+1 < n$. Our task is to show that the claim holds for $j > i$. Consider $d_q E_j / d_q x_i = E_j^{d_q x_i}$. Since E is defined from $\mathbb{R} \to \mathbb{R}$, the delta derivative construction reduces to the traditional case. Also, by definition, $E_j(x_j, \ldots, x_1) = E_{j+1}(f_{j+1}, x_j, \ldots, x_1)$.

Therefore,

$$
\begin{aligned}
E_j^{d_q x_i} &= E_{j+1}^{d_q x_i} \\
&= \frac{d_q E_{j+1}}{d_q x_1}\frac{d_q x_1}{d_q x_i} + \frac{d_q E_{j+1}}{d_q x_2}\frac{d_q x_2}{d_q x_i} \cdots \\
&\qquad + \frac{d_q E_{j+1}}{d_q x_j}\frac{d_q x_j}{d_q x_i} \\
&\qquad + \frac{d_q E_{j+1}}{d_q f_{j+1}}\frac{d_q f_{j+1}}{d_q x_i}
\end{aligned}
\tag{6.34}
$$

for $i \leq j$. From our definition from the preliminaries applied to our recursive definition of the ordered variables x_i, we have that $d_q x_k / d_q x_i = 0$ when $k < i$, as the preceeding variables in the order are unaffected by the later variables in the causation chain. This result allows us to reduce our equation to $E_j^{d_q x_i} = \frac{d_q E_{j+1}}{d_q x_i} + \frac{d_q E_{j+1}}{d_q f_{j+1}}\frac{d_q f_{j+1}}{d_q x_i}$. We collapse the first remaining term so that it matches the form of our induction hypothesis $E_{j+1}^{d_q x_i} = \sum_{k=j+2}^{n} x_n^{d_q^+ x_k} f_k^{d_q x_i}$, giving us

$$
\begin{aligned}
E_j^{d_q x_i} &= \sum_{k=j+2}^{n} x_n^{d_q^+ x_k} f_k^{d_q x_i} + \frac{d_q E_{j+1}}{d_q f_{j+1}}\frac{d_q f_{j+1}}{d_q x_i} \\
&= \sum_{k=j+1}^{n} x_n^{d_q^+ x_k} f_k^{d_q x_i}
\end{aligned}
\tag{6.35}
$$

which is our desired result. ∎

Thus, the chain rule for ordered derivatives in the quantum calculus is established. With this result, neural network architectures in the quantum calculus can be constructed and trained via backpropagation. While the traditional chain rule of classical analysis fails to hold for ordered derivatives, the chain rule for ordered derivatives does hold on q-time scales. Since time scales in general, and q-time scales in particular, may be the appropriate mathematical framework to discuss a certain class of resource allocation problems, and dynamic programming concerns itself with optimization of multi-stage decision scenarios, a quantum calculus approach to the approximation of the optimal solution becomes an exciting new area of computational decision theory.

6.6 Conclusions

Ordered derivatives and the backpropagation update rule have been established using the emerging mathematical field of time scales calculus. This calculus unifies the discrete and continuous domains, so our results provide a complete theoretical framework for discussing learning in connectionist systems that can admit input signals of any type.

Chapter 7
Unified Computational Intelligence in Social Science

7.1 Introduction

Having already discussed the use of unified computational intelligence to *learn* and to *adapt*, this book now investigates its ability to *seek*. Computational social science modeling allows heightened understanding of the dynamics of complex systems in ways that the traditional analytical approaches could not. In this way, unified computational intelligence algorithms can power models unlike anything computable using a static or mathematical approach. Agent-based modeling, using agents whose intelligence includes full-blown creativity thanks to their ability to *learn* and to *adapt*, is revealing information about ourselves and the world that has never before been supported. From how elephants mourn their dead to how pandemics spread to large-scale financial market models, these techniques are giving humanity a way to *seek* that used to be only the purview of mystics and philosophers. In domains where the unified approaches to learning and adapting prove advantageous, their combined ability to assist in seeking may be great.

As the study of agent-based computational social science grows, so does the need for appropriate techniques for the modeling of complex dynamic systems and the intelligence of the constructive agent. This chapter frames the problem and provides examples for which the unified computational intelligence techniques described in previous chapters can be used.

This chapter forms the third part of this book. It contributes an awareness of what these computational models can achieve. Too often, researchers in one specialty are unaware of progress being made in other areas. Humanity's ability to compute is a universal good; it is not to be filed away within a Computer Science or Computer Engineering curriculum and never seen by those pursuing the B.A., M.A., or EdD. degrees. The hope of this part of the book is that those specializing in both computation and social science or the humanities will be made aware of the existence of other fields and of how synergistic their union could be. Computing is not just for those who taught themselves hexadecimal in seventh grade, nor are the social sciences and the humanities just for those who cannot handle mathematics or write a nested loop. Everyone is in this together, and this chapter is a call for unity.

J. Seiffertt & D.C. Wunsch: Unified Computational Intell. for Complex Sys., ALO 6, pp. 91–109.
springerlink.com

Section 7.2 comes from a graduate-level course on Computational Intelligence and Game Theory taught at Missouri University of Science and Technology in the spring semester of 2009. Cross-listed in the Computer Engineering, Mathematics, Economics, and Computer Science departments, this course, designed by the author from scratch, represents the author's diverse background and brought together computation, analysis, and social science in a way that armed the students with the tools to perform research in these areas. This section discusses where computational intelligence fits into the picture as far as game theory and social modeling is concerned.

Section 7.3 is adapted from material that appeared as a chapter in a book on neural networks in economics and business (Zhang, 2007). It details the agent-based economics literature and presents ways for unified computational intelligence architectures to contribute in a fundamental way to advances in this research area.

Section 7.4 appeared in the IEEE Computational Intelligence Magazine in a special issue on computational finance. As the world struggles to make sense of recent financial crises, due in part to failures in mathematical modeling, the time for those trained in computation to become aware of the opportunities in areas beyond the scope of traditional engineering is upon us. Major advances await in the understanding of financial markets, and unified computational intelligence architectures can help build the models to achieve such advances.

7.2 Game Theory and Computational Social Science

Everything evolves. The world is made up of complex systems, from society to ecosystems to economies to our own brains. All are connected, and these connections are vastly more complicated than traditional analytic methods have been able to explain. To understand these systems at the heart of the world around us, it is time to turn to new tools made possible by advances in technology. The language of evolutionary game theory, complemented with the power of computational intelligence, holds the promise to guide us towards uncovering a sort of dynamic previously unknown.

This section seeks to study this still emerging approach to modeling, which relies on computational techniques, specifically agent-based computational models, instead of traditional dynamical system models used in the past to describe the complexity of massive interaction. To this end, we will explore the agent-based techniques of the field of computational intelligence and use them in tandem with the explanatory ability of traditional game theory.

7.2.1 Computational Intelligence

To find a local extrema of a differentiable function, a traditional analytic approach such as Newton's Method will proceed according to a fixed rule. First, choose an initial guess $\boldsymbol{x_0}$ and then update according to

$$x_{n+1} = x_n + \frac{f'(x_n)}{f''(x_n)} \tag{7.1}$$

If a solid starting point has been chosen, then this algorithm is sure to deliver the desired optimal value. This algorithm does not adapt. It simply is. Compare to a basic agent-based computational intelligence algorithm that generates a population of agents x_0^i and then send them off to optimize via

$$x_{n+1}^i = x_n^i + r_1\left(\operatorname*{argmax}_{n,i} x_n^i - x_n^i\right) + r_2\left(\operatorname*{argmax}_{n} x_n^i\right) \tag{7.2}$$

At first glance, this may look similar to Newton's Method. However, further inspection reveals two important differences common in computational intelligence techniques.

First are the random numbers r_1 and r_2. These numbers add an exploratory aspect to the algorithm not found in Newton's Method. These random numbers may lead the agents in directions a straight gradient descent would never dream of visiting. In fact, this element can be viewed as representing a level of *creativity* that gives this algorithm a much better chance of avoiding getting lost in suboptimal basins of attraction such as local extrema if the initial guess falls within certain regions which are, of course, impossible to know beforehand. It is for this reason that the most popular method of training neural networks, called backpropagation and based on a cousin of Newton's Method, is slowly being cast aside in favor of the methods of computational intelligence. When seeking a global optima, it is desirable to use a method that is less likely to succumb to the temptations inherent in a simple update rule such as Newton's Method. Instead, some bit of randomness is used to allow the agents making up the algorithm to *adapt* their way out of suboptimal choices.

The second important difference between these two optimization schemes is found in the interplay of the terms $\operatorname{argmax}_{n,i} x_n^i$ and $\operatorname{argmax}_n x_n^i$. The former indicates the most optimal position found by any agent thus far in the simulation, and the latter records the individual agent's best effort. These maxes, particularly the former, provide a way for the agents to *interact* with each other during the optimization run. This interaction, while stemming from simple basic rules, can lead to massively unpredictable aggregate emergent behavior within the entire society of agents. In fact, the term $\operatorname{argmax}_{n,i} x_n^i$ is called the *social* component, as it forces the agents to keep in mind what their leader is doing. It also introduces an *exploitation* tendency into the system to counter the *exploration* instinct embedded in the stochastic influence of the variables r_1 and r_2. Without the social element, the algorithm can readily devolve into random search. It is in the interplay between these two pulls on the agent's will that the true complexity begins to emerge. This algorithm can be modified further by adding a term to represent a neighborhood effect in which each agent is moved by the standing of the other agents closest to it, momentum terms that weigh the various influences in different ways, or specialist agents or entire coevolutionary societies all interacting in assorted ways and generating quite the array of complex emergent behavior.

It is also worth noting that the agent-based approach requires no knowledge of the function itself, nor does it constrain the nature of the function in the slightest. Whereas Newton's Method not only requires that the function be twice differentiable but also that it be possible to calculate these derivatives, the computational intelligence algorithm can attack any function regardless of smoothness and without knowing anything about it aside from the points evaluated during the simulation. Also, it is not hindered by inflection points.

These two elements—randomness and social interaction—are hallmarks of computational intelligence techniques. Genetic algorithms rely on random selection and the makeup of other agents in the society. Even neural networks involve individual neurons interacting in ways that are ordered yet unspecified in the aggregate. While basic feedforward networks may be adequately modeled analytically, the networks of the recurrent variety and those with complex and heterogeneous transfer functions very quickly stress the explanatory ability of the classical mathematics. Neural network function approximators go far beyond standard statistical regression techniques in a way similar to the manner in which agent-based methods shine more brightly than a seventeenth century invention such as Newton's Method.

This idea of complex behavior emerging from simple rules is a powerful one. It is the secret to *adaptation* and is the reason agent-based models of complex systems enjoy a character distinct from those of traditional analysis. While at some level it is true that an algorithm such as the agent-based optimizer presented above can be cast as a highly coupled stochastic dynamical system and approached from the direction of traditional analysis, such abstraction only serves to capture the simplest of agent-based methods. Even then, such analysis, while helpful to a degree, fails to reveal anything decisive about the algorithm that cannot be observed from the computation itself. Indeed, another trait found in many agent-based methods is imperviousness to the mathematician's formal proofs. The mathematician can only watch as the agents swarm towards the desired extrema; she cannot prove mathematically that they *must* do so, even given unlimited time to act or some other equally implausible simplifying assumption common in this genre of philosophy.

All of these gains from agent-based models are being realized due to massive increases in computing power. Scientists no longer have to write equations to model systems of interest. Instead, they can design robust simulations in which to test hypotheses and draw conclusions about the world. This *science of simulation* is made possible by the ability to carry high-powered computing devices in a pocket. This technological advancement has allowed computationally armed investigators to go beyond ink and papyrus, beyond ballpoint pen and paper, and even beyond stylus and touchscreen as modeling tools. The parameters of this young science have not yet been established, and while much thrashing is still occurring due to its nascence, it is already showing promise and making important contributions in many areas.

To some, the fact that this science's theory has yet to be penetrated to a meaningful extent by the tools of mathematical logic and formal proof is a dagger through its heart. How, they ask, can you use an optimization algorithm without knowing with certainty that the result you get is the one desired? How can an engineering system built using this technology be reliable? These questions ignore two vital points: (1) the assumptions needed for technical precision more often than not are of such an impractical character as to not be of use outside a mathematics journal and (2) bridges were built way before humans had a solid theory of statics and dynamics with which to analyze their construction. This is not to say that proof-centric results have no value; just the opposite, in fact. They have tremendous value, and computational scientists who fail to grasp the nuances of mathematical rigor are selling their work short. However, the same scientists need not wait for the mathematicians to prove that an agent-based optimizer converges according to some abstract optimality criterion in some excessively long period of time before applying the tool to actual problems at hand. Neither the mathematician nor the engineer has primacy. They are connected in a wild sort of dance in which the responsibility for leading is constantly up in the air. Engineers plow forward and develop techniques that work in practice. In turn, mathematicians take these roughshod ideas, add some shine, and craft a reflection of humanity's experience in the world consisting of pure form and essence. Then the applied folks read this literature and find inspiration for new ideas of their own. The fact that agent-based computational science has yet to admit the sort of coherent structure found in dynamical systems theory is a feature, not a bug. The next time a mathematician alludes to this, kindly remind him or her that Newton invented and used the derivative two centuries before it was rigorously defined. Computational science is following in the footsteps of the calculus, proving itself in the field before being properly and fully understood and defined.

Why the word *intelligence*? What is meant by computational *intelligence*, exactly? How can a computer be *intelligent*? After all, is it not said that computers can only do what you tell them to do? How is that *intelligent*? Rather than getting caught up in a discussion of a precise meaning of *intelligence*, it is more productive to consider what it is that these algorithms produce. Yes, it is factually true that computers only do what they are programmed to do. However, this statement is also misleading. Computers can, after all, be programmed to write a program (or even to construct a proof!) that a human thinker could not have come up with on his own. Through the use of adaptive learning techniques, researchers have taught computers to play Backgammon at a level surpassing that of any human player. In fact, these computational methods have produced strategies that were new to human experts, and this is a game that humanity has played and studied for five-thousand years. If a human player is capable of intelligence and creativity, perhaps these same qualities can be attributed in a nontrivial way to the play of a machine capable of exhibiting such characteristics. If games remain unconvincing, then how about poetry, art, and music? All of these have been created at a high level using adaptive agents. Whatever the fruit of *intelligence* may be, it is clear that the production generated by this level of computation is at least in the running to be ranked among such. In the end, many people would characterize intelligence as the ability to *adapt* to a changing and complex environment in the pursuit of some goal and, as mentioned earlier, it is this quality of *adaptation* that forms our central theme.

The ancients looked at the stars and invented mathematics to describe their patterns. These mathematical techniques have come a long way from their origins in prehistory, but they still have their limits. Humans now desire to look beyond the movement of stars and into their very hearts to see how they work and breathe, live and die. To do this, sophisticated partial differential equation models are written that capture the nuances observed by astrophysicists. However, when it comes time to solve these equations, to really get a feel for the development of stars, these equations are not solved analytically. Rather, computational models provide the greatest insight. Humanity has gone from a people who write about the movement of stars in philosophy to a people who discover how the stars are built using technology. It is humanity's captive attention to the stars above that has guided so much of history. Maybe the astrologers are onto something.

7.2.2 Agent-Based Computational Social Science

Traditionally, scientists studying the mechanisms of human exchange would approach their investigations using the tools bequeathed to them by colleagues enamored of particles, energy and, yes, stars. Importantly, the mathematical frameworks produced in the nineteenth century by Leon Walras stand today as the technology most used by economists to describe the complexities of exchange, scarcity, and markets. These general equilibrium models seek to characterize the interplay of prices, demand, and supply. To use them for such, economists incorporate common simplifying assumptions to produce tractable mathematics. Many empirical studies question these assumptions, and new schools of economic thought are being developed to address their shortcomings. The field of Agent-Based Computational Economics is one of these schools. Drawing on advancements in our technology, it is able to analyze complex systems in a way unknown to nineteenth and even to some extent twentieth century minds. It is also a call of sanity, as social science researchers need not toil under the same mathematical stressors that drove Georg Cantor to madness!

Researchers in this field are able to ask new questions. In particular, the Walrasian equilibrium models assumed the existence of an oracle that set prices for the entire economy. However, human history shows us that in the absence of central control, societies tend to develop regularities. Recently developed *generative* approaches use agents with flexibility of action to study the emergence of global consistencies. Public policy analysts are interested in valuing a host of competing policy plans. Computational models provide a laboratory of sorts for the introduction of a variety of rules, the results of which can then be gathered as evidence in either support of or opposition to a given plan. This addresses the general disadvantage that social scientists face when running experiments as compared to their peers in physics and chemistry labs. When natural experiments—that is, simply waiting for events to occur on their own and then rushing to gather data—prove uncooperative, computational economies may be of great assistance. Also, these artificial economies can be rerun time and time again with different initial conditions or assumptions. In this way, a study of global dynamics can be undertaken that would have been unthinkable in Walras's time. Also, as in the above discussion of the computational intelligence approach to optimization, computational agents have an ability to interact with each other that is unsupported in traditional approaches. Agents can send each other messages in an adaptive, unscripted way, either to simply say "this point right here is the best I've found so far" as in the optimizer, or in more elaborate socially and economically meaningful ways. Finally, perhaps the greatest feather in agent-based computational economics' digital cap is its ability to usefully investigate *out-of-equilibrium* phenomena. The classical general equilibrium models only tell us what price levels lead to markets clearing. In the real world, however, out-of-equilibrium systems are observed on a continual basis, as exogenous shocks occur with regularity and unpredictability. The mathematics of dynamical systems, born as it is from the study of the continuous motion of objects through Newtonian space, is insufficient for the theoretical study of these sorts of system states. Computationally, however, the entire life of the economy can be witnessed, from inception through various equilibria and basins of attraction.

Economics is not the only social science benefiting from agent-based modeling. Even in purely social systems, the idea of explaining a pattern is nearly equivalent to showing that it is the equilibrium state of some analytic model. To study how a system attains equilibrium or to discuss out-of-equilibrium behavior, other methods are required. In particular, an agent-based model of the Anasazi civilization shows how computational models can be used by archaeologists and anthropologists. This model is able to generate a society, fit to actual archaeological and geographical data, and provide an explanation of the society's history based on the actions of individuals rather than relying on models that require the homogeneity of agents and that are constrained so as to permit closed-form analytic solutions.

Social scientist and Brookings Institute senior fellow Joshua Epstein even argues for a new standard for scientific explanation: "If you didn't grow it, you didn't explain it." He feels that agent-based methods, or *generative* methods, are in fact *the* tool to be used in scientific investigation.

This all said, and the power of equilibrium models having been questioned, attention is now directed to a field that has at its core the definition of, and search for, equilibria.

7.2.3 Game Theory

It is perhaps not a coincidence that it was Jon von Neumann, computing pioneer extraordinaire, who, along with economist Oskar Morgenstern, wrote the first text on game theory and with it the first analytical approach to discussing social interaction in a formal way. The focus of this section, after all, is the study of ways in which computational techniques have aided this study.

Beginning as a formal description of competitive two-person strategy in parlor games, game theory has evolved into a general framework for both competitive and cooperative dynamic interaction among any number of agents over any conceivable time horizon. A more appropriate name would be Multi-Agent Interaction Theory, but the term "game" seems here to stay and represents any sort of strategic interaction among various agents, or players. Game theory wields much explanatory power even in its basic analytic form, even considering its heavy reliance on the calculation of equilibria. For example, the Talmud presents a classic problem about division of resources to creditors after one's death. It gives three examples, two of which were easily understood but the third of which remained a mystery to scholars for literally thousands of years. It wasn't until the 1980s that the division rules were fully understood using the tools of game theory.

This chapter first will explain the basic mathematical structure of game theory, including normal and extensive form games and equilibria. It is important to have an appreciation for rigor when doing analytical modeling work, and a significant contribution game theory has made to those looking to discuss profitably the nuances of social interaction is a precise language. Topics covered will include Nash equilibria, dominated strategies, credible and incredible threats, and other terms carrying clear technical weight. These terms help to stage the basic interplay of agents working to achieve some goal while considering the actions of others. Determining what solution is optimal is a major component of game theory, and this chapter will cover many ways to calculate equilibria.

We will then move on to the theory of repeated and evolutionary games. When an agent is placed in a situation in which a given encounter must be undertaken repeatedly, a different sort of equilibrium strategy emerges. These evolutionary dynamics will be discussed in detail and are a particularly harmonious fit for the adaptive flair of computational intelligence techniques. Trigger strategies, the Folk theorems, reputation effects, the replicator dynamics, and evolutionary stability will be among the topics covered.

Finally, some time will be spent discussing behavioral games and agent-based approaches to social modeling. Behavior game theory challenges dearly held economic assumptions regarding the nature of rationality by running actual experiments with human subjects and comparing the results with the predicted equilibria generated by hyper-rational agents. Agent-based approaches, discussed in more detail in the previous section, will act as a capstone. They stand at the frontier of social modeling, drawing upon the power of the computer and the spirit of game theory's rigor.

7.3 Economics and Finance

7.3.1 Introduction

Economists have long recognized their inability to run controlled experiments *a la* their physicist and biologist peers. As a result, while much real science can be done using natural experiments, analytic mathematical modeling, and statistical analysis, a certain class of discoveries regarding the governing dynamics of economic and financial systems has remained beyond the grasp of such research. However, recent advances in computing show promise to change all that by gifting economists with the power to model large-scale, agent-based environments in such a way that interesting insight into the underlying properties of such systems can be obtained. It is becoming increasingly evident that engineering tools from the area of computational intelligence can be used in this effort.

Agent-based methods are enjoying increased attention from researchers working in economics as well as in pure and applied computation. The central focus of this still nascent field involves the generation of populations of interacting agents and the observation of the resulting dynamics as compared to some optimality criterion, analytically or otherwise obtained. Typically, some sort of learning algorithm, such as a simple feed-forward, multi-layer perceptron neural network, will be implemented in the model. Often, other techniques of computational intelligence, such as genetic algorithms, will be used to evolve the population, showing the promise that gains in this area of computation have for social science investigation.

7.3.2 Background

The fundamental Agent-Based Computational Economics framework structure is overviewed in Testafasion (2006) and will be reviewed here. The particular formulation of the agent problem proposed in this chapter is based on the presentation in Chiarella (2003) and will be discussed following the general overview. Finally, other supporting literature will be surveyed to help solidify the main ideas of this section and to guide the reader in other directions of possible research interest.

7.3.3 Agent-Based Computational Economics

A standard course of study in economics grounds the reader in a host of equilibrium models: the consumer preference theory of microeconomics (Binger, 1998), the wage determination cycle of labor economics (Ehrenberg, 2003), the concept of purchasing power parity in international finance (Melvin, 2000), and the Walrasian Auctioneer (Leijonhufud, 1967) of macroeconomics. In all of these approaches to describing economic phenomena, the student is presented with top-down analytic treatments of the dynamics of an entire economy's worth of individual interacting agents. While the local scale behavior informs the higher-level dynamics, it is only the global portion that enjoys specific elucidation. Exactly how the lives of the agents respond to an economic shock in order to return the system to the long-run equilibrium is not considered. Furthermore, the level of simplifying assumptions necessary to achieve clear and acceptable results from an analytical model, via some fixed-point theorem, often serves to cast a significant measure of doubt over the entire affair. Importantly, this problem is not a fixture of economics alone; these models and the chase for mathematically provable periodicity results permeates other areas of science, notably population biology (Bohner, 2006). Also, many proof-theoretic approaches require overly restrictive and wholly unrealistic linearity assumptions to arrive at a tractable model, denying insight that claims the answer to an economic question may have more than one root cause (Judd, 2006.)

The discipline of Agent-Based Computational Economics (ACE) analyzes an economy from another point of view, one termed "constructive" due to the focus on the fundamental elements of the system as opposed to the global dynamics. Instead of specifying long-run equilibrium behavior, the ACE researcher takes care to capture in his or her equations the salient behaviors of the individual agents. Any emergent dynamics or long-run convergence will reveal themselves as a result of the collection of individual choices. In this way, equilibrium models can be tested in a manner akin to a controlled experiment in the physical sciences. The population of computational agents can be constrained in a certain way, and the resulting dynamics explored via simulation. Such studies can confirm theoretical fixed-point, long-term equilibrium results or serve as evidence that such hallowed equations may be missing something quite vital about the system's reality.

For example, Hayward (2005) finds that computational experimental modeling fails to support standard analytic models for price forecasting and trading strategies in international financial markets. He finds, in contradiction to the notion that a trader's success is a function of risk aversion instead of proficiency in accurate forecasting, that the agents with short time horizons in an environment with recurrent shocks emerge as dominant, as they adapt to and learn about the nature of the economic system in which they operate. His work incorporates genetic algorithms and multi-layer perceptron neural networks which, along with swarm intelligence and fuzzy logic methods, are core areas of the computational intelligence field (Engelbrecht, 2002).

ACE models begin by specifying attributes and modes of interaction among the agents. One way to implement this specification is through an object-oriented programming approach, wherein the agents could be considered objects, the attributes private data members, and modes of interaction publicly-accessible methods. The books by Johnsonbaugh (2000) and Horstmann (2004) include details on object-oriented programming, the specifics of which are not integral to our current discussion. Another tool accessible to researchers conducting ACE investigations is one of the standardized modeling frameworks, such as the one published by Meyer (2003). Finally, analytic equation models can be found in the literature, such as the early work of Lettau (1997). It should be noted that these models, while analytic in nature, still conform to the constructive ACE philosophy in that they are employed in the characterization of the salient features of the agents. The equations are not being used to set the dynamics of the system a priori or to launch a search for a periodic equilibrium solution.

Whatever agent representation a researcher chooses, it is important that the computational intelligence technique used to model the agent's ability to adapt to a complex environment be sufficiently robust to generate accurate and substantive results. An experiment that seemingly shows a population of agents unable to learn to converge to an analytic equilibrium may not really be unearthing a new economic truth; instead, this could be an indication that the particular computational learning algorithm employed in the simulation is insufficient for the complexity of the task. Furthermore, care must be taken to appropriately read the output of an ACE simulation. Unlike standard econometric approaches (Greene, 2003 and Kennedy, 2001), it is often difficult to calculate a level of statistical confidence to accompany the conclusions of an ACE model. It should be noted here that the computational techniques falling under the banners of Adaptive Resonance architectures and Partially Observable Markov Decision Processes, discussed later in this chapter, have the advantage that they come equipped with readily available confidence level information, thus assuaging this objection to numerical investigation of economic phenomena. In any case, an increase in knowledge of advanced computational techniques, such as those in computational intelligence, will go a long way towards overcoming the inertia naturally present in any community in the face of change in paradigm, as better communication and pedagogy will help to ward off the feeling among many that these algorithms are simply "black boxes" akin to a foul sorcerer's magic that should not be trusted.

While Hayward (2005) used a multi-layer perceptron architecture to model how the agents learned to project financial information, more robust results may be gained by using sophisticated time series prediction techniques (Cai, 2004, Hu, 2004) or the techniques overviewed later in this chapter.

7.3.4 Application to Economic Systems

Computational economic agents must think. Their entire raison d'etre is to provide researchers with guidance in addressing questions about the governing laws of dynamic systems. To extract the most value from the ACE approach, the most advanced computational tools available should be considered.

It is critical that the computational agent be able to effectively process information within the environment. Consider the formulation of Chiarella (2003). They construct a population of agents engaged in the decision of whether to buy or sell a particular share of an asset. The economy consists of two assets: a risky asset with price Pt and dividend dt, and a risk-free asset with known rate of return r for every epoch t. The agents model using a benefit function $V_{it,}$ encapsulating their understanding of the market at a given point in time. This study involves heterogeneous agents, so one group of agents uses a market signal to calculate this V_{it} and another group pays a cost c_t for access to the theoretical fundamental solution $F_t = \sum_{i=1}^{\infty}(1+r)^{-1}E_t(d_t)$, which is the summation of discounted future expected dividends. An approach to this problem type using ADP and Adaptive Critics is a natural extension of the existing work. Furthermore, these techniques will allow investigation into more complex, higher-scale systems. In particular, it is important to consider these techniques when faced with a highly nonlinear complex system, such as a large-scale economy or financial market.

Following the work of Duffy (2006) on comparison to controlled economic experiments using human subjects, researchers need to accurately model the agent's cognitive processes as they apply to economic activity. The ART family of neural network architectures (Carpenter and Grossberg, 1991) is ideally suited to such a task, given its roots in the mathematical modeling of the operation of the human brain.

It is an exciting time to be involved in computational economics and finance. The advances in computational intelligence techniques bring quite a bit of promise to the investigation of some basic but major problems of emergent system dynamics.

7.3.5 Future Research Directions

Much work must be done to expand ADP techniques to other application areas, particularly in an operations research setting, where the tremendous scale of industrial-scale logistics problems pushes the limits of current computational power. Theoretical developments need to be chased that can address this problem, as it is not sufficient to wait for the computer architects to design next-generation processor capabilities. The scaling issue that these algorithms face as the dimensionality of the problem increases is a major stumbling block.

As pointed out in Young (2006), agent-based methods are important for studying many sorts of emergent social phenomena, including the emergence of money as the outcome of an iterated coordination game. Other social dynamics can be studied and progress made towards their understanding using these techniques. This level of human discernment can have a great positive impact on all our lives, beyond the realm of a financial market environment or mathematical psychology.

While researchers currently employ techniques such as genetic algorithms and multi-layer perceptron neural networks, considerable room for growth exists by using more advanced approaches. As these techniques become more widely understood, they may shed their image as a "black box" (LeBaron, 2006). Approximate Dynamic Programming, influenced so heavily by the economic strategic risk assessment literature, is particularly well suited for widespread application as the computational force behind agent thinking processes.

Finally, these are research technologies capable of bringing together communities of researchers from seemingly disparate fields to approach a wide range of important problems. Much good can come from such a globalized approach to collaborative investigation.

7.4 Intelligence in Markets

The study of multi-agent market interactions is becoming increasingly dependent upon the concurrent development of appropriate computational tools. The field of computational intelligence is ideally positioned to offer much to this new wave of financial and economic science. This section introduces aspects of two types of market interaction: an agent's decision to distribute resources among assets and manipulating the structure of the market to influence the nature of the interaction of the agents within. Natural computation techniques such as artificial neural networks, genetic algorithms, Approximate Dynamic Programming, and particle swarm optimization are heavily used among researchers in these growing fields. With further attention, it is possible for engineers to make great strides in both understanding these complex systems and in building commercial applications for them.

7.4.1 Introduction

Evolution, human and animal cognition, and the emergent coordination of systems of autonomous agents are among the areas of nature drawn upon for inspiration by the field of computational intelligence. Researchers are increasingly using these aspects of nature in the exploration of market interaction, both to develop more scientific knowledge about economic and financial phenomena and to build new commercial applications for price forecasting, asset trading, and market design. Adam Smith wrote of an "invisible hand" that guided markets. Today, scientists may consider this specter to be the result of a limiting process of some computational evolutionary algorithm. It is here, at the intersection of economics and intelligent computation, where the most profitable study of the nature of human market interaction may occur.

Markets provide a way to convey information regarding prices of assets. For those working within a market, the goal is to most optimally allocate a pool of assets to satisfy some predetermined optimality criterion. For those outside the market, the goal is to optimally design the allocation mechanisms to achieve the specified mission of the central planners. Together, these two processes can be studied within the larger view of the role of markets in complex systems as a form of communication and coexistence. Computational intelligence, taking its cues in part from nature, has much to offer the calculation aspects of these two concepts. This section discusses the application of asset allocation and mechanism design using elements of natural computation.

Markets are widely considered the most efficient method of processing transactions. There is a wealth of economic literature speaking to the efficacy of free markets under certain, rather strenuous constraints that support these ideas. However, it is also useful to study the behavior of markets outside these constraints. Mechanism design permits an analysis of market institutions away from those used in efficiency theorems with the goal of developing new efficient trading structures.

Natural computation can be seen in the forefront of a number of research areas. An important example is that of intelligent control, which has plentiful application in economics and finance. The field of Approximate Dynamic Programming (Si, Barto, Powell, & Wunsch, 2004) offers hope that many of the previously intractable control problems can be handled. Partial motivation for its baseline techniques comes from studies of the human mind. Though neuroscientists do not yet fully understand how the human brain works, the study of its processes is of increasingly fundamental importance for applied engineers and economists.

Economists, unable to run the actual controlled experiments common in the physical sciences, have turned to natural experiments, statistical investigation, and mathematical analytic modeling to understand their science. The new paradigm of ACE gives economists the ability to model large-scale, agent-based environments in order to make new discoveries regarding the governing dynamics of the economic and financial systems under review. The tools of computational intelligence are becoming increasingly relevant to this new field.

The fundamental feature of this emerging field of economics requires generating populations of interacting agents followed by an analysis of the resulting dynamics. These agents are then used to test the claims of equilibrium theorems. It is common to see some element of computational intelligence, such as a simple feed-forward, multi-layer perceptron neural network or a genetic algorithm, used as a learning rule or overlying constraint within the model.

ACE approaches analyze an economy from an angle different from that of the analytical models. It is called a "constructive" approach accounting for its focus on the fundamental elements of the system (e.g., the individual agents) as opposed to global dynamics. Any emergent dynamics or long-run convergence is determined by the aggregation of individual choices. In this way, analytic equilibrium models can be tested by generating a population of computational agents and exploring the resulting dynamics. Researchers can then assess theoretical predictions using the simulations.

For example, Hayward (2005) uses ACE techniques to show evidence that accepted analytic models governing the forecasting of prices and trading strategies in international financial markets fail to adequately predict the behavior of computational agents. His findings show that traders adapt to the nature of the market environment rather than relying solely on accurate forecasting. His work incorporates genetic algorithms and multi-layer perceptron neural networks.

This section presents an overview of recent uses of computational intelligence techniques in asset pricing and mechanism design applications. Additionally, it discusses ways to extend these results through the use of even more robust methods of natural computation. The aim is to give computational intelligence researchers an idea of the scope of work being pursued in the areas of asset pricing and mechanism. Many opportunities for commercial application exist in these financial fields for those with an interest in applying complex intelligent systems to markets.

7.4.2 Approximate Dynamic Programming and Stochastic Control

Dynamic Programming, developed by Richard Bellman in the middle of the twentieth century, is perhaps the most correct way to think about multi-stage decision problem solving. It gives rise to the Bellman equation, which aids in the calculation of optimal strategies. This equation takes the form of the Hamilton-Jacobi equation of classical mechanics and is thus oftentimes labeled the Hamilton-Jacobi-Bellman equation in the literature. It is interesting to note the similarities between the trajectories of decisions made by agents in a multi-stage problem environment and the motion of point-like particles through space. Much research focuses on this equation, and the discipline of ADP, mentioned above, seeks approximations of this equation sufficient for handling significant industrial-scale, real-time control problems.

The mathematics of dynamic programming has also been utilized in the asset pricing literature. In Ben-Amour, Breton, Agouti, & L'Ecuyer, 2007, these techniques are used to price options embedded in bonds. It is important for practitioners to discern the proper valuation of these assets and for researchers to report that their dynamic programming approaches outperform techniques common historically in mathematics, such as finite-difference methods.

While their analysis takes place within a well-structured artificial market, it remains a control problem at heart. The computational intelligence techniques of ADP, which are capable of generating solutions to the dynamic programming problem in complex, changing environments, may be useful in taking this approach to a level in which it could be used by asset managers in the field.

ADP approaches typically utilize a neural network as a function approximator within their design. For simpler problems, the neural network can be replaced with a lookup table housing all the values of a given state-action pairing. The idea, then, is for the agent to look at the table for the given state and choose the action containing the highest value. This algorithm is called Q-learning and is the focus of work in Lee & Park, 2007. In their study, the stochastic control problem of stock trading is addressed. They deploy a system of agents, all learning via the Q-learning heuristic, who tackle the problem of profitable trading. Measuring performance in both the profit and risk management dimensions, they report that this computational intelligence approach outperforms alternative approaches.

It is notable that their approach mixes in a sense both the ACE approach and Q-learning. A system of agents, working cooperatively, provides an emergent intelligence that achieves quality performance in the stock trading task. Such results indicate that much further good can come from a more systematic application of both basic and advanced techniques from the computational intelligence literature into problems being studied within the ACE framework. It may indeed turn out that an understanding of human market systems is reliant on our development of more robust algorithms inspired from our natural environments.

Stochastic control, involving varying degrees of uncertainty, is a prime target for ADP and other computational intelligence approaches. In Song & Huang, 2007, this paradigm is leveraged in a setting with discrete decision points. The trading agent is to make decisions in a Markov Decision Process (MDP) environment with stochastic returns on investment while taking into account transactions costs. MDP problems, particularly those of the partially observable variety (POMDP's), have received much attention from the computational intelligence community. The results presented in this study show that research into POMDP's, and the fact that ADP handles the unobservability criterion, may show promise in the commercial application of computational intelligence techniques to trading decisions within a market.

7.4.3 Evolving Asset Pricing Strategies

Stock market trading algorithms are a favorite topic of forecasters and financial researchers. In Bekiros, 2007, a neurofuzzy approach is studied. Taking a cue from cognitive modeling of the human brain and combining it with a mathematical apparatus capable of handling uncertainty when dealing with imprecise information, this paper shows the hybrid computational intelligence method outperforming both a simple recurrent neural network and basic statistical regression models.

While this research focuses on technical trading rules, it is encouraging to believe that further advances along this direction may lead to more robust results. Typically, asset managers take into account fundamentals, technical signals, and automated data, and the development of an intelligent system that can include more of the data used in practice would be a welcome commercial application.

It is worth noting that the idea of asset allocation, and the mathematics underlying the study of the topic within financial markets, is not limited to modeling only the trading behavior of those dealing in stocks, options, and bonds. It can also be generalized to apply to any environment in which choices need to be made. In Rice, McDonnell, Spydell, & Stremler, 2006, the assets in question are the targets and threats in an air strike simulation. Evolutionary computation is employed to generate a player for a simulation in which a player must defend his own assets from attack as well as divert resources to destroy the assets of the opponent. The same asset allocation and pricing machinery developed for financial markets, including the use of dynamic programming, can be utilized profitably in this environment as well.

Extensions of financial market concepts to areas of interest outside the realm of economics are consequences of the nature of the underlying mathematics. Just like the Bellman equation of decision theory holds abstract similarity to the Hamilton-Jacobi theory describing the motion of particles in classical physics, so can the basic approach to asset allocation in markets be applied to other areas of engineering interest. So, even for engineers with limited interest in economics and finance, the study of these approaches may be fruitful.

Algorithms based on observed evolution in natural species have played a prominent role in recent economics studies. In Matilla-Garcia, 2006, genetic algorithms are used to evolve trading rules for computational market players. The author compares the genetic algorithm designed trader to a buy-and-hold strategy and concludes that the computational intelligence boosted agent outperforms the other. Additionally, simulations are run comparing the genetic algorithm-based strategies within differing market conditions: bullish, bearish, and volatile.

The ability of computational intelligence techniques to handle complex environments and aggregate disparate data leads to these advances. As trading in markets is a form of the intelligent control problem, it is not surprising that these approaches are well suited to the commercial task.

Options submit to the well-known Black-Scholes formula for pricing. Itself a deft application of stochastic partial differential equations to the modeling of economic phenomena, it is possible to go beyond this analytic formalism using the algorithms of computational intelligence. In Zapart, 2003, neural networks are used to push past the constraints of the Black-Scholes equation and price these complicated assets appropriately.

Neural networks and binomial trees are one approach presented in this work as a volatility model. The feature space of volatility is represented as wavelets that feed the neural network predictions. The paper also develops genetic algorithm-based neural network architectures capable of reproducing the Black-Scholes results. Given that analytical approaches such as Black-Scholes rely heavily on unrealistic simplifying constraints or come attached to system requirements proving inapplicable in commercial application, the fact that these computational intelligence techniques can provide the same or better return as the traditionally accepted analytic methods is heartening.

Fuzzy logic and evolutionary processes are used in Ghandar, Michalewicz, Schmidt, To, & Zurbruegg, 2007, to calculate trading rules for equity markets. This work is intriguing because, in contrast to most approaches in this area, it is rule based. The genetic algorithm uses chromosomes that represent fuzzy rules capable of parsing ambiguities such as "high volatility" and "extremely high volume." Discussions with human traders reveal that they think along these lines, giving rules of thumb based on such imprecise language. Compared to a technical rule that posits a sell or buy given a fixed numerical value for volatility or volume, these fuzzy rule-based systems represent to a higher degree the way human cognition operates.

One approach to asset allocation is to control the value-at-risk of a given portfolio. As a control problem, this is well suited to ADP and other computational intelligence techniques. In Chapados & Bengio (2001), a neural network approach is taken. One of their approaches combines a neural network forecast with traditional mean-variance portfolio optimization theory. The computational intelligence engine provides an understanding of the price of the assets, and then the efficient frontier is searched for the optimal portfolio based on comfortable risk levels. This is more prediction than true asset allocation, but it does demonstrate the value that natural computation can add to these processes.

The second approach taken is more in the line of intelligent control as it calculates buy-sell decisions based on computational intelligence techniques. Recurrent neural networks power a decision engine designed to minimize transaction costs and optimize a cost function representing financial performance. The network outputs a vector of recommendations, and the recurrent nature of the system allows for past recommendations to be considered when making future buy-sell decisions. Together, this system and the previously mentioned neural network forecasting architecture outperform benchmark market performance. Of course, managers in the field may hesitate to use the signals generated from such machine learning algorithms, but as the computational intelligence community continues to develop dependable trading systems, these attitudes may indeed relax. Commercial application possibilities for these sorts of systems are wide open and available to anyone with a solid system and a marketing plan.

The mean-variance approach to asset allocation is studied in Dashti, Farjami, Vedadi, & Anisseh, 2007, as a swarm optimization application. The authors take the modern portfolio theoretic approach of optimizing a basket of assets while taking into account the risk factors associated with each one. Using the Sharpe ratio as the fitness level for the swarm, the authors evolve optimal portfolios using a population-based optimizer rather than using the more traditional analytical approach. The advantages here lie in being able to optimize over more robust solution spaces, which may indeed provide a better modeling framework for the sorts of markets mangers find in practice.

Population-based optimization algorithms are perhaps ideally suited to investigations of market action in the ACE paradigm. Since what is of interest is, fundamentally, the nature of the emergent behavior of a number of interacting agents, the internal structure of an algorithm such as Particle Swarm Optimization allows the researcher to not only simply optimize a given fitness function but to do it in a way that is representationally meaningful in terms of their simulation. In fact, further complexity can be introduced by considering the individuals in these optimizers as the economic agents themselves. This actually touches on an analytic modeling approach such as overlapping generations models, which have provided economists with much value already.

7.4.4 The Design of Market Mechanisms

Mechanism design is applied to electric power markets in Silva, Wollenberg, & Zheng, 2001. The authors posit a need for a certain type of regulation of the electric power market, as the unregulated market may not provide incentives for the necessary sharing of information in order to perform required economic dispatch of the generation. Their work, then, focuses on how best to structure the market to meet the specific needs of electric power generation. The electric power market is of a different character than other markets where free competition leads to optimal equilibria. The existence of transmission systems with uncontrollable flows of power may lead to congestion or even to overloading of the system. Therefore, some regulation is required, and the question is how best to design the constraints to maximize desired variables.

With a model representing the transmission network constraints, they apply mechanism design to produce optimality when each participating firm acts in its best interest. They note that the mechanism they develop is different from those in the existing literature, thus showcasing the vitality of the game theoretic approach. Their result is that the market should be organized in a bidding process whereby each firm presents a production cost level that another agent then uses in assigning electricity production and payments. It is worth observing that the application of mechanism design to an engineering market was able to produce an offering that was not evident to human planners organizing these markets in practice. This tool of mechanism design, particularly when augmented with computational intelligence, has much to offer central planners in any number of sectors.

To study auction and market designs, researchers in the ACE vein have introduced a number of artificial markets and agent structures. One, called ZIP (zero-intelligence-plus), is studied in Cliff, Walia, & Byrde, 2003. Combining mechanism design and genetic algorithms, the authors develop optimal auction rules for their given environment. Furthermore, as is often the case when using evolutionary computation, the results are far different from those put in place by human planners.

Instead of copying existing auction mechanisms, the authors were able to evolve hybrid mechanisms. The ZIP traders adapt to auction conditions in order to optimize some internal criterion. Other research fixed the auction mechanism in advance and studied the emergent behavior of the individual trading agents. The impact of this study is that they allowed the auction mechanism itself to adapt. The authors created a population of 30 individuals representing a 9-dimensional vector describing the mechanisms. Fitness evaluation was undertaken by measuring the emerging market dynamics. The best mechanism design was then evolved in this fashion.

These are particularly interesting results on account of the fact that they can be seen as an invitation for computational intelligence methods to inform the coordination of market actors in non-auction environments as well.

7.4.5 Computational Markets

The new approach to treating markets as emergent results of computational algorithms is important in economics. The review Mirowski, 2007, covers much of this new paradigm. Researchers and practitioners in computational intelligence have the ability to stand at the forefront of this new movement and to make substantial contributions to the scientific understanding of market phenomena as well as to the development of new positive commercial applications.

References

Abramson, M., Audet, C.: Convergence of mesh adaptive direct search to second-order stationary points. SIAM Journal of Optimization 17(2), 606–619 (2006)

Ahlbrandt, C., Bohner, M., Ridenhour, J.: Hamiltonian systems on time scales. J. Math. Anal. Appl. 578, 250–561 (2000)

Ahlbrandt, C., Morian, C.: Partial differential equations on time scales. Journal of Computational and Applied Mathematics 141(1-2), 35–55 (2002)

Alpaydin, E.: Introduction to machine learning, pp. 1–3. MIT Press, Cambridge (2004)

Al-Timini, A., Abu-Khala, M., Lewis, F.: Adaptive critic designs for discrete-time zero-sum games with application to H-infinity control. IEEE Transactions on Systems, Man, and Cybernetics 1(37), 240–247 (2007)

Antinolfi, G., Azariadis, C., Bullard, J.: Monetary policy as equilibrium selection. Federal Reserve Bank of St Louis Review 89(4), 331–341 (2007)

Amis, G., Carpenter, G.: Default ARTMAP 2. In: Proceedings of the International Joint Conference on Neural Networks, Orlando, FL (2007)

Arifovic, J.: Genetic algorithm learning and the cobweb model. Journal of Economic Dynamics and Control 18, 3–28 (1994)

Arifovic, J.: The behavior of the exchange rate in the genetic algorithm and experimental economies. Journal of Political Economy 104, 510–541 (1996)

Arrow, K.J.: Historical Background. In: Arrow, K., Karlin, S., Scarf, H. (eds.) Studies in the Mathematical Theory of Inventory and Production. Stanford University Press, Stanford (1958)

Arthur, W.B.: Inductive reasoning and bounded rationality. American Economic Review 84, 406–411 (1994)

Arthur, W.B., Holland, J., LeBaron, B., Palmer, R., Tayler, P.: Asset pricing under endogenous expectations in an artificial stock market. In: Arthur, W.B., Durlauf, S., Lane, D. (eds.) The Economy as an Evolving Complex System II, pp. 15–44. Addison-Wesley, Reading (1997)

Atici, F.M., Biles, D.C., Lebedinsky, A.: An application of time scales to economics. Mathematical and Computer Modelling 43(7-8), 718–726 (2006)

Aviv, Y., Pazgal, A.: A Partially Observable Markov Decision Process for Dynamic Pricing. Management Science 51(9), 1400–1416 (2005)

Bardi, M., Capuzzo-Dolcetta, I.: Optimal Control and Viscosity Solutions of Hamilton-Jacobi-Bellman Equations. Birkhauser, Boston (1997)

Basar, T., Olsder, G.: Dynamic Noncooperative Game Theory, Second Edition, 2nd edn. SIAM, Philadelphia (1999)

Baylor University Time Scales Group. Ti.me.. Sc.a...les MATLAB Toolbox v. 1.1 (2008), `http://www.timescales.org` (retrieved July 2, 2008)

Bekiros, S.: A neurofuzzy model for stock market trading. Applied Economics Letters 14, 53–57 (2007)

Bellman, R.: Dynamic Programming. Princeton University Press, Princeton (1957)

Bellman, R., Dreyfus, S.: Applied Dynamic Programming. Princeton University Press, Princeton (1962)

Beltratti, A., Margarita, S., Terna, P.: Neural Networks for Economic and Financial Modeling. International Thomson Computer Press, London (1996)

Ben-Ameur, H., Breton, M., Karouti, L., L'Ecuyer, P.: A dynamic programming approach for pricing options embedded in bonds. Journal of Economic Dynamics and Control 31, 2212–2233 (2007)

Bertsekas, D.: Dynamic Programming and Optimal Control, 2nd edn., vol. 1, 2. Athena Scientific, Belmont (2000)

Bertsekas, D., Tsitsiklis, J.: Neuro-Dynamic Programming. Athena Scientific, Belmont (1996)

Binger, B., Hoffman, E.: Microeconomics with Calculus. Addison-Wesley, Reading (1998)

Bohner, M.: Calculus of variations on time scales. Dynamic Systems and Applications 13, 339–349 (2004)

Bohner, M., Fan, M., Zhang, J.: Periodicity of scalar dynamic equations and applications to population models. Journal of Mathematical Analysis and Applications 330(1), 1–9 (2007)

Bohner, M., Hudson, T.: Euler-type boundary value problems in quantum calculus. International Journal of Applied Mathematics and Statistics 9(J07), 19–23 (2007)

Bohner, M., Guseinov, G.: Double integral calculus of variations on time scales. Computers and Mathematics with Applications 54(1), 46–57 (2007)

Bohner, M., Guseinov, G.: Multiple integration on time scales. Dynamic Systems and Applications 14(3-4), 579–606 (2005)

Bohner, M., Guseinov, G.: Partial Differentiation on Time Scales. Dynamic Systems and Applications 13, 351–379 (2004)

Bohner, M., Peterson, A.: Advances in Dynamic Equations on Time Scales. Birkhäuser, Boston (2003)

Bohner, M., Peterson, A.: Dynamic Equations on Time Scales: An Introduction with Applications. Birkhäuser, Boston (2001)

Brannon, N., Conrad, G., Draelos, T., Seiffertt, J.: Wunsch. D. Information fusion and situation awareness using ARTMAP and partially observable Markov Decision Processes. In: Proceedings of the IEEE International Joint Conference on Neural Networks, pp. 2023–2030 (2006)

Brannan, N., Seiffertt, J., Draelos, T., Wunsch, D.: Coordinated machine learning for situation awareness. Neural Networks 22(3) (2009)

Bullard, J., Seiffertt, J.: Japanese deflation loses something in the translation. National Economic Trends (September 2003)

Carpenter, G., Grossberg, S.: The ART of adaptive pattern recognition by a self-organizing neural network. Computer 21(3), 77–87 (1988)

Carpenter, G., Grossberg, S. (eds.): Pattern Recognition by Self-Organizing Neural Networks. The MIT Press, Cambridge (1991)

Carpenter, G., Grossberg, S., Reynolds, J.: ARTMAP: Supervised Real-Time Learning and Classification of Nonstationary Data by a Self-Organizing Neural Network. Neural Networks 4, 565–588 (1991)

Carpenter, G., Grossberg, S., Rosen, D.: Fuzzy ART: Fast Stable Learning and Categorization of Analog Patterns by an Adaptive Resonance System. Neural Networks 4, 759–771 (1991)

Carpenter, G., Grossberg, S., Markuzon, N., Reynolds, J., Rosen, D.: Fuzzy ARTMAP: A neural network architecture for incremental supervised learning of analog multidimensional maps. IEEE Transactions on Neural Networks 3(5), 698–713 (1992)

Carpenter, G., Markuzon, N.: ARTMAP-IC and medical diagnosis: Instance counting and inconsistent cases. Neural Networks 11, 323–336 (1998)

Cai, X., Zang, N., Venayagamoorthy, G., Wunsch, D.: Time series prediction with recurrent neural networks using a hybrid PSO-EA algorithm. In: Proceedings of the International Conference on Neural Networks, vol. 2, pp. 1647–1652 (2004)

Castro, J., Georgiopoulos, M., Secretan, R., DeMara, R., Anagnostopoulos, G., Gonzalez, J.: Parallelization of Fuzzy ARTMAP to Improve its Convergence Speed. Nonlinear Analysis: Theory, Methods, and Applications 60(8) (2005)

Chapados, N., Bengio, Y.: Cost functions and model combination for VaR-based asset allocation using neural networks. IEEE Transactions on Neural Networks 12(4), 890–906 (2001)

Chiarella, C., Gallegati, M., Leombruni, R., Palestrini, A.: Asset Price Dynamics among Heterogeneous Interacting Agents. Computational Economics 22, 213–223 (2003)

Clerc, M., Kennedy, J.: The particle swarm–explosion, stability, and convergence in a multidimensional complex space. IEEE Transactions on Evolutionary Computation 6(1), 58–73 (2002)

Cliff, D., Walia, V., Byde, A.: Evolved hybrid auction mechanisms in non-ZIP trader marketplaces. In: Proceedings of the IEEE Conference on CIFE, pp. 167–174 (2003)

DaCunha, J.: Instability results for slowly time varying linear dynamic systems on time scales. J. Math. Anal. Appl. 328, 1278–1289 (2007)

DaCunha, J.: Stability for time varying linear dynamic systems on time scales. Journal of Computational and Applied Mathematics 176, 381–410 (2005)

DaCunha, J., Davis, J., Singh, P.: Existence results for singular three point boundary value problems on time scales. J. Math. Anal. Appl. 295, 378–391 (2004)

Dashti, M., Farjami, Y., Vedadi, A., Anisseh, M.: Implementation of particle swarm optimization in construction of optimal risky portfolios. In: Proceedings of the IEEE Conference on IEEM, pp. 812–816 (2007)

Dietterich, T.: An experimental comparison of three methods for constructing ensembles of decision trees: Bagging, boosting, and randomization. Machine Learning 40(2), 139–157 (2000)

Denn, M.: Optimization by Variational Methods. McGraw-Hill, New York (1969)

Duarte, M., Hu, Y.: Vehicle classification in distributed sensor networks. Journal of Parallel and Distributed Computing 64(7), 826–838 (2004)

Duffie, D.: Dynamic Asset Pricing Theory, 3rd edn. Princeton University Press, Princeton (2001)

Duffy, J.: Agent-based models and human subject experiments. In: Testafasion, L., Judd, K. (eds.) Handbook of Computational Economics, vol. 2, pp. 949–1012. Elsevier, Amsterdam (2006)

Ehrenberg, R.G., Smith, R.S.: Modern Labor Economics. Theory and Public Policy. Addison-Wesley, Reading (2003)

Eloe, P., Hilger, S., Sheng, Q.: A qualitative analysis on nonconstant graininess of the adaptive grid via time scales. Rocky Mountain J. Math. 36, 115–133 (2006)

Endsley, M.: Toward a theory of situation awareness. Human Factors 37(1), 32–64 (1995)

Englebrecht, A.: Computational Intellligence: An Introduction. John Wiley, Chichester (2002)

Enns, R., Si, J.: Helicopter trimming and tracking control using direct neural dynamic programming. IEEE Transactions on Neural Networks 14(4), 929–939 (2003)

Evans, G., Honkapohja, S.: Learning and Expectations in Macroeconomics. Princeton University Press, Princeton (2001)

Fakih, S., Das, T.: LEAD: a methodology for learning efficient approaches to medical diagnostics. IEEE Transactions on Information Technology in Biomedicine 1(55), 158–170 (2006)

Ferreira, R., Torres, D.: Higher order calculus of variations on time scales. In: Proceedings of the Workshop on Mathematical Control Theory (2007)

Ferreira, R., Torres, D.: Remarks on the calculus of variations on time scales. Int. J. Econ. Ecol. Stat. 9, 65–73 (2007)

Fleming, W., Rishel, R.: Deterministic and Stochastic Optimal Control. Springer, New York (1975)

Gelfand, I.M., Fomin, S.V.: Calculus of Variations. Dover, Mineola (2000)

Ghandar, A., Michalewicz, Z., Schmidt, M., To, T., Zurbruegg, R.: A computational intelligence portfolio construction system for equity market trading. In: Proceedings of the IEEE Congress on Evolutionary Computation, pp. 798–805 (2007)

Gravagne, I., Marks, R., Davis, J., DaCunha, J.: Application of time scales to real time communications networks. In: American Mathematical Society Western Section meeting, special session on Time Scales, University of Southern California (2004)

Gravagne, I., Davis, J., Marks, R.: How deterministic must a real-time controller be? In: Proceedings of 2005 IEEE/RSJ International Conference on Intelligent Robots and Systems, Alberta, Canada, August 2-6, pp. 3856–3861 (2005)

Gravagne, I., Davis, J., DaCunha, J., Marks, R.: Bandwidth reduction for controller area networks using adaptive sampling. In: Proc. Int. Conf. Robotics and Automation, New Orleans, LA, April 2004, pp. 5250–5255 (2004)

Greene, W.: Econometric Analysis. Prentice-Hall, Englewood Cliffs (2003)

Grossberg, S.: Adaptive pattern classification and universal recoding. Biological Cybernetics 23, 187–202 (1976)

Guseinov, G.: Integration on time scales. J. Math. Anal. Appl 285, 107–127 (2003)

Guseinov, G., Ozyilmaz, E.: Tangent lines of generalized regular curves parametrized by time scales. Turkish Journal of Math. 25(4), 553–562 (2001)

Hall, D., Llinas, J.: An introduction to multisensor data fusion. Proceedings of the IEEE 85(1), 6–23 (1997)

Han, D., Balakrishnan, S.: State-constrained agile missile control with adaptive-critic based neural networks. IEEE Transactions on Control Systems Technology 10(4), 481–489 (2002)

Hayward, S.: The role of heterogeneous agents' past and forward time horizons in formulating computational models. Computational Economics 25(1-2), 25–40 (2005)

Hilger, S.: Ein Maßkettenkalkül mit Anwendung auf Zentrumsmannigfaltigkeiten. PhD Thesis, Universität Würzburg (1988)

Hilger, S.: Analysis on measure chains—a unified approach to continuous and discrete calculus. Results Math. 18, 18–56 (1990)

Hilscher, R., Zeidan, V.: Calculus of variations on time scales: weak local piecewise C_{rd}^1 solutions with variable endpoints. J. Math. Anal. Apppl. 289(1), 143–166 (2004)

Hortsmann, C.: Object-Oriented Design and Patterns. Wiley, Chichester (2004)

Hu, X., Wunsch, D.: Time series prediction with a weighted bidirectional multi-stream extended Kalman filter. In: Proceedings of the IEEE International Joint Conference on Neural Networks, vol. 2, pp. 1641–1645 (2004)
Hull, D.: Optimal Control Theory for Applications. Springer, New York (2003)
Iyer, M., Wunsch, D.: Dynamic re-optimization of a fed-batch fermentorusing adaptive critic designs. IEEE Transactions on Neural Networks 12(6), 1433–1444 (2001)
Jackson, B.: Partial dynamic equations on time scales. Journal of Computational and Applied Mathematics 186, 391–415 (2006)
Javaherian, H., Liu, D., Zhang, Y., Kovalenko, O.: Adaptive critic learning techniques for automotive engine control. In: Proceedings of the American Control Conference, vol. 5, pp. 4066–4071 (2003)
Johnsonbaugh, R., Kalin, M.: Object-Oriented Programming in C++. Prentice-Hall, Englewood Cliffs (2000)
Judd, K.: Computationally intensive analysis in economics. In: Testafasion, L., Judd, K. (eds.) Handbook of Computational Economics, vol. 2, pp. 882–893. Elsevier, Amsterdam (2006)
Kac, V., Pokman, C.: Quantum Calculus. Springer, New York (2001)
Kelly, E., Kennedy, P.: A dynamic stochastic model of mate desertion. Ecology (74), 351–366 (1993)
Kennedy, J., Eberhart, R.: Particle swarm optimization. In: Proceedings of the IEEE International Conference on Neural Networks, pp. 1942–1948 (1995)
Kennedy, P.: A Guide to Econometrics. MIT Press, Cambridge (2001)
Kokar, M., Tomasik, T., Weyman, J.: Formalizing classes of information fusion systems. Information Fusion 5(3), 189–202 (2004)
Kulkarni, N., Krishna, K.: Intelligent engine control using an adaptive critic. IEEE Transactions on Control Systems Technology 11(2), 164–173 (2003)
LeBaron, B.: Agent-based computational finance: Suggested readings and early research. Journal of Economic Dynamics and Control 24, 679–702 (2000)
LeBaron, B.: Agent Based Computational Finance. In: Testafasion, L., Judd, K. (eds.) Handbook of Computational Economics, vol. 2, pp. 1187–1235. Elsevier, Amsterdam (2006)
Lee, J.W., Park, J.: A multiagent approach to Q-learning for daily stock trading. IEEE Transactions on Systems, Man, and Cybernetics—Part A: Systems and Humans 37(6), 864–877 (2007)
Leijonhufud, A.: Keynes and the Keynesians: A suggested interperatation. American Economic Review 57(2), 401–410 (1967)
Lettau, M.: Explaining the facts with adaptive agents: The case of mutual fund flows. Journal of Economic Dynamics and Control (21), 1117–1148 (1997)
Lewis, F., Syrmos, V.: Optimal control, 2nd edn. Wiley, New York (1995)
Lin, C.: Adaptive critic autopilot design of Bank-to-turn missiles using fuzzy basis function networks. IEEE Transactions on Systems, Man, and Cybernetics 35(2), 197–207 (2005)
Neely, C., Weller, P., Dittmar, R.: Is technical analysis in the foreign exchange market profitable? A genetic programming approach. Journal and Finance and Quantitative Analysis 32(4), 405–426 (1997)
Marks, R., Gravagne, I., Davis, J., DaCunha, J.: Nonregressivity in switched linear circuits and mechanical systems. Mathematical and Computer Modelling 43, 1383–1392 (2006)
Matilla-Garcia, M.: Are trading rules based on genetic algorithms profitable? Applied Economics Letters 13, 123–126 (2006)
Melvin, M.: International Money and Finance. Addison-Wesley, Reading (2000)

Messer, K.: Riccati techniques on a time scale. Panamer Math J. 13(2), 1–18 (2003)

Meyer, D., Karatzoglou, A., Leisch, F., Buchta, C., Hornik, K.: A Simulation Framework for Heterogeneous Agents. Computational Economics (22) (October-December 2003)

Miranda, M., Fackler, P.: Applied Computational Economics and Finance. MIT Press, Cambridge (2002)

Mirowski, P.: Markets come to bits: Evolution, computation, and markomata in economic science. Journal of Economic Behavior and Organization 63, 209–242 (2007)

Mohagheghi, S., del Valle, V., Venayagamoorthy, G., Harley, R.: A proportional-integrator type adaptive critic design-based neurocontroller for a static compensator in a multimachine power system. IEEE Transactions on Industrial Electronics 54(1), 86–96 (2007)

Mohagheghi, S., Venayagamoorthy, G., Harley, R.: Optimal wide area controller and state predictor for a power system. IEEE Transactions on Power Systems 22(2), 693–705 (2007)

Moore, B.: ART 1 and pattern clustering. In: Touretzky, D., Hinton, G., Sejnowski, T. (eds.) Proceedings of the 1988 Connectionist Models Summer School. Morgan Kauffman, San Francisco (1989)

Morrison, J., Kelly, R., Moore, R., Hutchins, S.: Tactical decision making under stress (TADMUS) decision support system. In: Proceedings IRIS National Symposium on Sensor and Data Fusion. MIT Lincoln Laboratory, Lexington (1997)

Muchoney, D., Williamson, J.: A Gaussian adaptive resonance theory neural network classification algorithm applied to supervised land cover mapping using multitemporal vegetation index data. IEEE Transactions on Geoscience and Remote Sensing 39(9), 1969–1977 (2001)

Muttel, C.: The Black Scholes Equation in Quantum Calculus (2007)

Potzsche, C.: Chain rule and invariance principle on measure chains. J. Comput Appl. Math. 141, 249–254 (2002)

Powell, W.: Approximate Dynamic Programming: Solving the Curses of Dimensionality. Wiley Series in Probability and Statistics, Hoboken (2007)

Principe, J., Euliano, N., Lefebvre, W.: Neural and Adaptive Systems. Fundamentals Through Simulations. John Wiley & Sons Inc., New York (2000)

Padhi, R., Balakrishnan, S.: Optimal management of beaver population using a reduced-order distributed parameter model and single network adaptive critics. IEEE Transactions on Control Systems Technology 14(4), 628–640 (2006)

Paul, J.: Smart Sensor Web: Web-based Exploitation of Sensor Fusion for Visualization of the Tactical Battlefield. IEEE AESS Systems Magazine (2001)

Prokhorov, D., Wunsch, D.: Adaptive Critic Designs. IEEE Transactions on Neural Networks 8(5), 997–1007 (1997)

Puterman, M.: Markov Decision Processes. Discrete Stochastic Dynamic Programming. Wiley Series in Probability and Mathematical Statistics, New York (1994)

Rice, A., McDonnell, J., Spydell, A., Stremler, S.: A player for tactical air strike games using evolutionary computation. In: Proceedings of the IEEE Symposium on Computational Intelligence and Games, pp. 83–89 (2006)

Rogers, J., Sheng, Q.: Notes on the diamond-α dynamic derivative on time scales. J. Math. Anal. Appl. 326, 228–241 (2007)

Routledge, B.: Genetic algorithm learning to choose and use information. Macroeconomic Dynamics 5, 303–325 (2001)

Sanyal, S.: Stochastic Dynamic Equations. PhD Dissertation. Missouri University of Science and Technology (2008)

Seiffertt, J.: An alpha derivative formulation of the Hamilton-Jacobi-Bellman equation of dynamic programming. In: IEEE International Joint Conference on Neural Networks, Atlanta, GA (2009)

Seiffertt, J., Sanyal, S., Wunsch, D.: Hamilton-Jacobi-Bellman equations and approximate dynamic programming on time scales. IEEE Transactions on Systems, Man, and Cybernetics, Part B 38(4), 918–923 (2008a)

Seiffertt, J., Sanyal, S., Wunsch, D.: Decision theory on dynamic domains: Nabla derivatives and the Hamilton-Jacobi-Bellman equation. In: IEEE Systems, Man, and Cybernetics Conference, Singapore (2008b)

Seiffertt, J., Wunsch, D.: Backpropagation and ordered derivatives in the time scales calculus. IEEE Transactions on Neural Networks (to appear)

Seiffertt, J., Wunsch, D.: A Single-ART Architecture for Unsupervised, Supervised, and Reinforcement Learning. In: Proceedings of the International Conference on Cognitive and Neural Systems, Boston, MA (2007)

Seiffertt, J., Wunsch, D.: Higher order neural network architectures for agent-based computational intelligence and finance. In: Zhang, M. (ed.) Artificial Higher Order Neural Networks for Economics and Business, IGI Global (2008)

Seiffertt, J., Wunsch, D.: A quantum calculus formulation of ordered derivatives and dynamic programming. In: Seiffertt, J., Wunsch, D. (eds.) IEEE International Joint Conference on Neural Networks, Hong Kong (2008)

Seiffertt, J., Wunsch, D.: Intelligence in markets: Asset pricing, mechanism design, and natural computation. IEEE Computational Intelligence Magazine 3(4) (2008)

Seiffertt, J., Van Brunt, A., Wunsch, D.: Maximum likelihood methods in biology revisited with tools of computational intelligence. In: IEEE Conference on Engineering in Medicine and Computational Biology, Vancouver, CN (2008)

Serrano-Gotarredona, T., Linares-Barranco, B.: A Low-Power Current Mode Fuzzy-ART Cell. IEEE Transactions on Neural Networks 17(6), 1666–1673 (2006)

Si, J., Barto, A., Powell, W., Wunsch, D.: Handbook of Learning and Approximate Dynamic Programming. IEEE Press Series on Computational Intelligence (2004)

Silva, C., Wollenberg, B., Zheng, C.: Application of mechanism design to electric power markets. IEEE Transactions on Power Systems 16(4), 862–869 (2001)

Sheng, Q., Fadag, M., Henderson, J., Davis, J.: An Exploration of Combined Dynamic Derivatives on Time Scales and Their Applications. Nonlinear Analysis: Real World Applications 7, 395–413 (2006)

Song, H., Huang, H.: Dynamic stochastic programming for asset allocation problem. In: Proceedings of the IEEE Conference on IEEM, pp. 16–20 (2007)

Spann, M., Vlassis, N.: Perseus: Randomized point-based value iteration for POMDP's. Journal of Artificial Intelligence Research 24, 195–220 (2005)

Stokey, N., Lucas, R., Prescott, E.: Recursive Methods in Economic Dynamics. Harvard University Press (1989)

Sutton, R.: TD Models: Modeling the world at a mixture of time scales. In: Prieditis, A., Russell, S. (eds.) Proceedings of the Twelfth International Conference on Machine Learning, pp. 531–539. Morgan Kaufmann, San Francisco (1995)

Sutton, R., Barto, A.: Reinforcement Learning. MIT Press, Cambridge (1998)

Tesauro, G.: TD-Gammon, a self-teaching backgammon program, achieves master-level play. Neural Computation 6(2), 215–219 (1994)

Testafasion, L., Judd, K. (eds.): Handbook of Computational Economics: Agent Based Computational Economics. North-Holland, Amsterdam (2006)

Todorov, E.: Linearly Solvable Markov Decision Problems. In: Proceedings of NIPS (2007)

Vasilic, S., Kezunovic, M.: Fuzzy ART neural network algorithm for classifying the power system faults. IEEE Transactions on Power Delivery 20(2), 1306–1314 (2005)

Watkins, C.: Learning from delayed rewards. PhD thesis. Cambridge University (1989)

Werbos, P.: Beyond Regression. PhD Dissertation. Harvard University (1974)

Werbos, P.: Backpropagation through time: What it does and how to do it. Proceedings of the IEEE 78(10) (1990)

Werbos, P.: Consistency of HDP applied to a simple reinforcement learning problem. Neural Networks 3(2), 179–189 (1990)

Werbos, P.: Neural networks and the human mind: new mathematics fits humanistic insight. In: Proceedings of the IEEE International Conference on Systems, Man, and Cybernetics, vol. 1, pp. 78–83 (1992)

Werbos, P.: The Roots of Backpropagation: From Ordered Derivatives to Neural Networks and Political Forecasting. Wiley, New York (1994)

Werbos, P.: ADP: Goals, Opportunities, and Principles. In: Si, J., Barto, A., Powell, W., Wunsch, D. (eds.) Handbook of Learning and Approximate Dynamic Programming. IEEE Press, Piscataway (2004)

Werbos, P.: Backwards differentiation in AD and neural nets: past links and new opportunities. In: Bucker, M., Corliss, G., Hovland, P., Naumann, U., Boyana, N. (eds.) Automatic differentiation: applications, theory, and implementations. Springer, New York (2005)

Williamson, J.: Gaussian ARTMAP: A Neural Network for Fast Incremental Learning of Noisy Multidimensional Maps. Neural Networks 9(5), 881–897 (1996)

White, D., Sofge, D. (eds.): Handbook of Intelligent Control. Van Nostrand (1992)

Wintz, N.: The Kalman Filter on Time Scales. PhD Dissertation. Department of Mathematics and Statistics, Missouri University of Science and Technology (2009)

Wunsch, D., Caudell, T., Capps, C., Marks, R., Falk, R.: An optoelectronic implementation of the adaptive resonance neural network. IEEE Transactions on Neural Networks 4(4), 673–684 (1993)

Xu, R., Anagnostopoulos, G.: Wunsch. D. Multiclass cancer classification using semisupervised ellipsoid ARTMAP and particle swarm optimization with gene expression data. IEEE/ACM Transactions on Computational Biology and Bioinformatics 4(1), 65–77 (2007)

Xu, R., Wunsch, D.: Clustering. IEEE/Wiley Press (2008)

Young, H.: Social dynamics: Theory and applications. In: Testafasion, L., Judd, K. (eds.) Handbook of Computational Economics, vol. 2, pp. 1082–1107. Elsevier, Amsterdam (2006)

Zapart, C.: Beyond Black-Scholes: a neural networks-based approach to options pricing. International Journal of Theoretical and Applied Finance 6(5), 469–489 (2003)

Zhang, M., Xu, S., Fulcher, J.: Neuron-Adaptive Higher Order Neural-Network Models for Automated Financial Data Modeling. IEEE Transactions on Neural Networks 13(1) (2002)

GPSR Compliance

The European Union's (EU) General Product Safety Regulation (GPSR) is a set of rules that requires consumer products to be safe and our obligations to ensure this.

If you have any concerns about our products, you can contact us on ProductSafety@springernature.com

In case Publisher is established outside the EU, the EU authorized representative is:

Springer Nature Customer Service Center GmbH
Europaplatz 3
69115 Heidelberg, Germany

Batch number: 10682290

Printed by Printforce, the Netherlands